Hydroelectric Sublime
水力美学

Hydroelectric Sublime
水力美学

Photographs
Beatrice Gorelli
Keiichi Kitayama

Texts
Nicolas Nova
Takako Itoh

Lars Müller Publishers

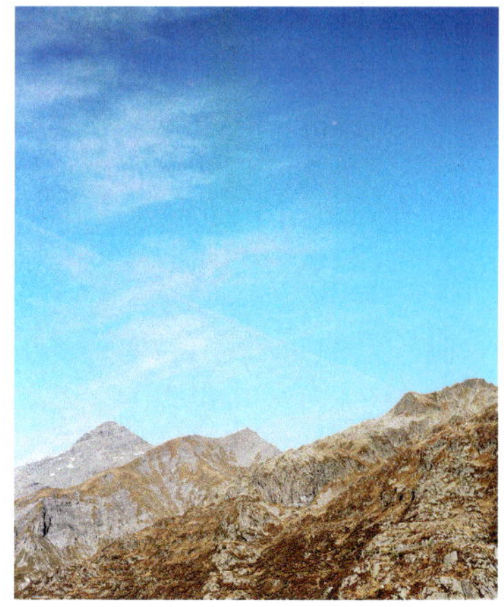

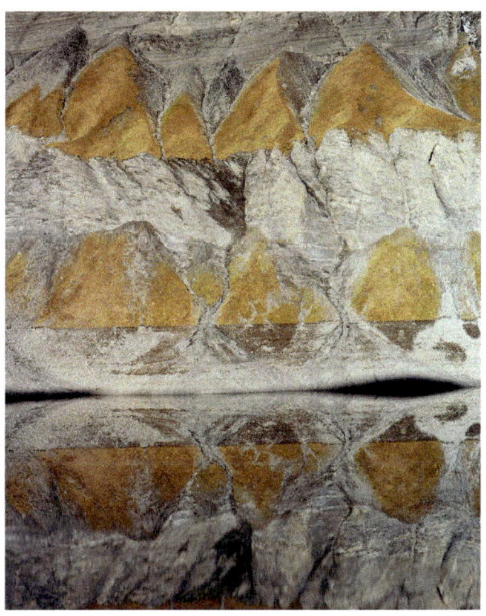

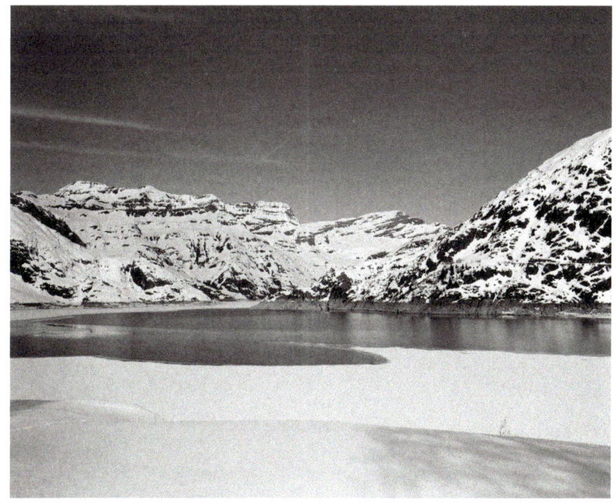

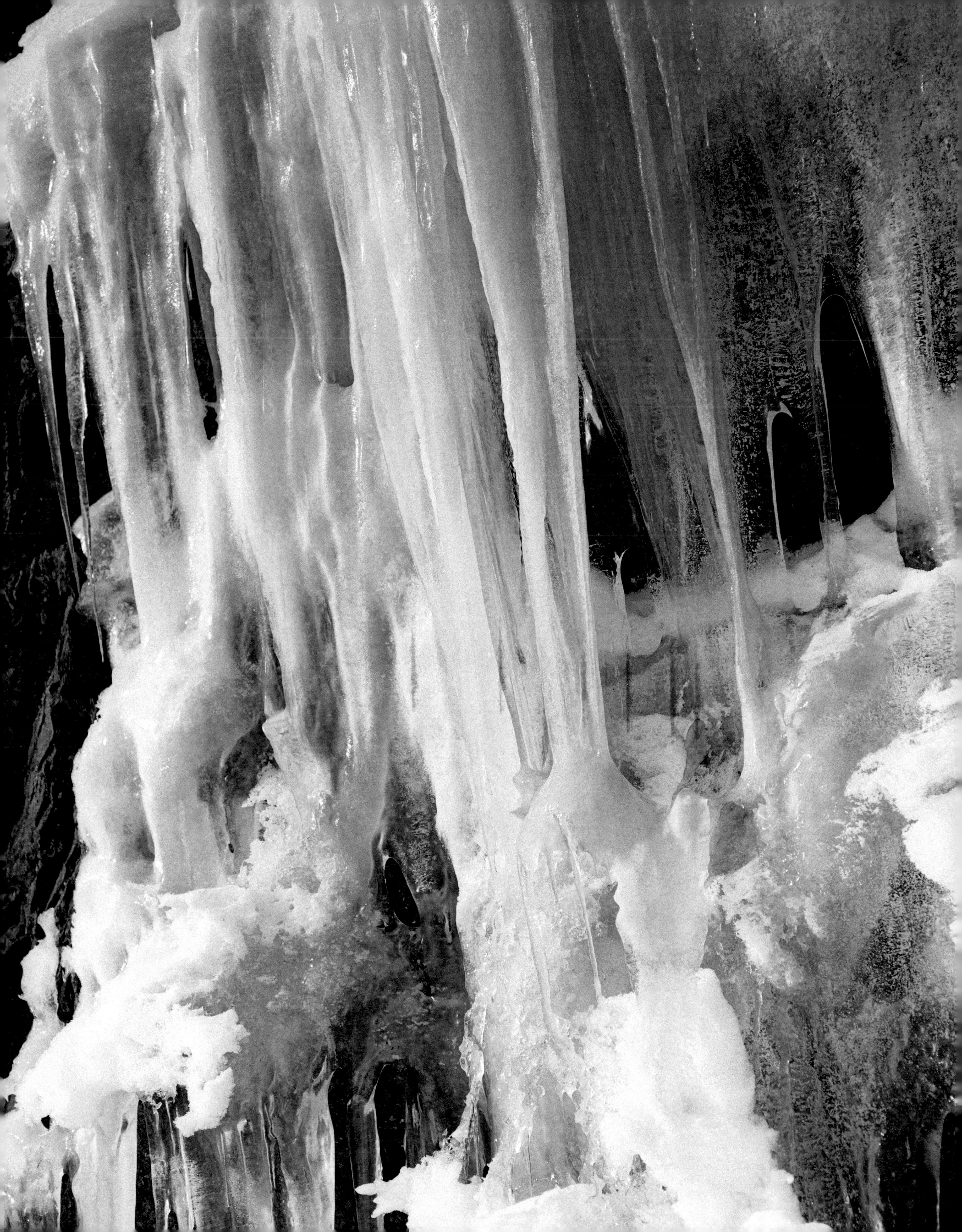

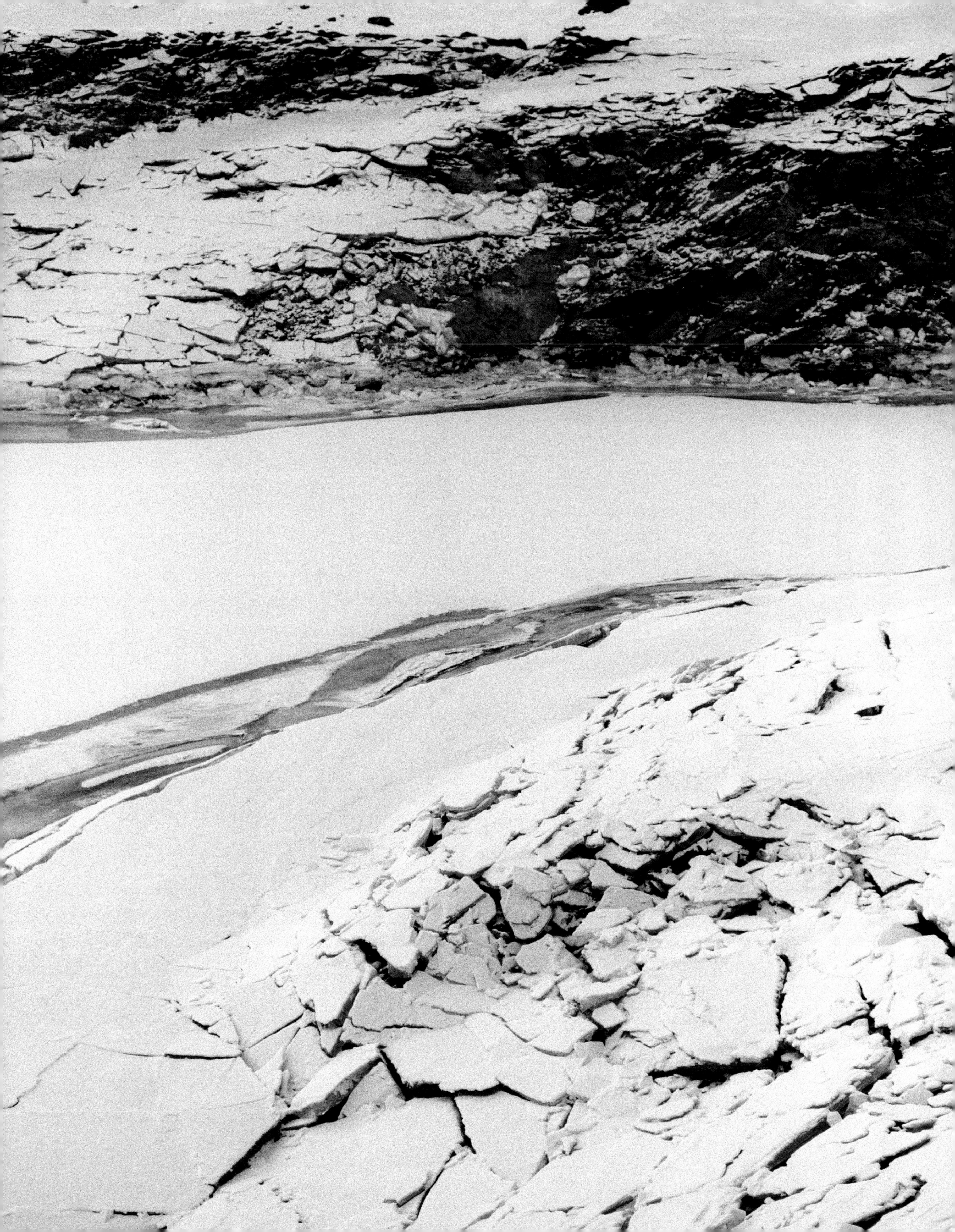

Water tower, electric cathedral, landscape machine, water guardian, concrete monster. The vocabulary used to designate hydroelectric dams is rich and extensive. And particularly in the Alps, where the dams' imposing appearance contrasts with the mineral context of the natural highland landscape in which they are nestled.

Depending on the viewpoint of the person using them, these expressions reveal different connotations—sometimes negative, sometimes emphatic—about the existence and appearance of such megastructures. The metaphors of the dam express a complex symbolism linked as much to the functional character of these places—to store energy or water—as to the architectural gesture they represent. In this respect they embody modern hubris, by their construction in a difficult environment, itself marked by the long history of Alpine geology, and also by the village societies they came to shake up, as described, for example, by the Valaisan writer Maurice Chappaz.[1] Beyond their direct usefulness, it is undoubtedly this monumental character that has progressively given these ensembles the roles of tourist attraction, leisure space, or place of memory.[2] And at the same time, to celebrate the power of human beings to dominate their environment, and perhaps also to remember the efforts and tragedies that their construction caused.

Borrowing from a culture marked by the networking of the world, I cannot help but find these metaphors highly local, and exclusively focused on the architecture of the place they seek to describe. As a twenty-first-century observer, accustomed to all sorts of communication systems, from telephone cables to the internet to geolocation systems, I consider these expressions as seeming to neglect the ownership and networked nature of these dams. One can, of course, understand how the rapid construction of these megastructures in the previous century provoked a shock requiring reference to previous devices that were just as monumental (the religious edifice, industrial machinery) or to deformed legendary creatures (the giant, the monster). But what strikes me when I observe an Alpine dam such as that of Emosson is their nature, and their insertion in a complex interlacing. This interweaving operates on multiple geographical scales: locally because of the structures and the cavities or pipes that link them together, regionally because of the link with the electricity production and transport infrastructure that are connected to it, and then, more distantly, by supplying energy to multiple faraway territories. The landscape of the dam is not strictly local; it exists within a network. In other words, we are dealing with a networked landscape.

1 Maurice Chappaz, *Chant de la Grande Dixence* (Lausanne: Payot, 1965; new edition by Éditions de L'Aire, 1995).

2 Tristan Loloum, "La vie touristique des grands barrages hydroélectriques," *Mondes du Tourisme* (2016), 12.

Because of this sprawling spatial presence, and most certainly because of its persistence over time, the Emosson dam constitute a particular figure of the nature/technology hybridization typical of the Anthropocene, the new geological epoch during which the influence of the human being on the Earth will reach such a level that it will become a major geological force. Documented here by Beatrice Gorelli and Keiichi Kitayama, it is this new figure that makes the place unique.

A remarkable example of hydroelectric engineering, the Emosson infrastructure forms a complex made up of three main dams: Emosson, Barberine, and Vieux-Emosson. They are all connected by tunnels and various cavities, and the complex is reinforced by the presence of the underground pumping station of the Nant de Drance, which is as discreet on the surface as it is gargantuan in depth.

Assembled from edifices that have been renewed several times since its construction, together these different elements form an integrated system for storing and producing hydroelectric energy on a large scale. This organization in the form of a local Franco-Swiss network is also manifested by the fact that water from the neighboring glaciers of Trient, Lognan, Le Tour, and Argentière is captured by canals and penstocks that carry it to the Emosson dam. This vast network of adductions, comprising nearly 40 kilometers of galleries, collects water from an area of 176 square kilometers to fill the 225 million cubic meters of the Emosson lake.

For the casual observer, such a local network is almost invisible. Apart from the arches testifying to the majestic presence of the dams, it's not easy initially to identify the signs of such a complex. The water collectors are discreet, the tunnels and canals are buried in the mountains, and one of the dams (Barberine) can only be seen when the level of the Emosson reservoir is very low. And if the periods of work, with the disruption that they cause in the valley, have testified to the changes underway in the Emosson complex, their completion have returned the places to the previous quietness of the mountains.

As a symbol of the tourist-attraction character of the dams, this local complex is also coupled with an infrastructure that allows you to visit and admire its monumental nature: the old funicular, built to transport the material and personnel necessary for the construction of the Barberine dam, has become a means of opening up access for tourists. The same goes for the mini-funicular that allows you to climb above the dam. Or the little panoramic train that winds through forests and tunnels so you can contemplate the Mont-Blanc massif from the mountainside route.

If the hydroelectric and transport complex forms an obvious local system, it must be remembered that it extends into a sprawling network. Firstly in the vicinity, via the water pipes that allow the generation of electricity in the two neighboring power stations of Vallorcine in Haute-Savoie (France) and La Bâtiaz in Martigny (Switzerland). These infrastructures are themselves connected to the electricity transmission network, for example by means of a very high voltage underground line at La Bâtiaz, or by more traditional overhead lines with its assembly of pylons and cables visible in the landscape, and then to various electricity transformation stations that ensure the supply to the population or to public and private organizations. By plugging in all our everyday objects, we connect ourselves without thinking too much about the landscape-network of the heights.

3 Thomas P. Hughes, "The Evolution of Large Technological Systems," in *The Social Construction of Technological Systems*, eds. E. Bijker, Th. P. Hughes, and T. Pinch, (Cambridge, Mass.: MIT Press, 1987).

4 Peter Haff, "Technology as a geological phenomenon: Implications for human well-being," *Geological Society, London, Special Publications,* 395 (2014), 301–309.

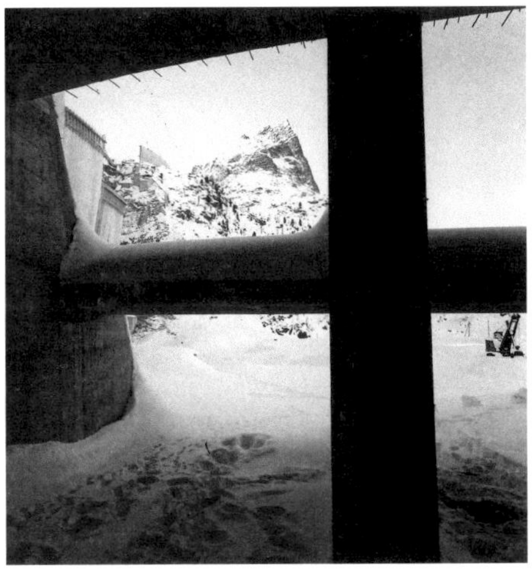

Emosson dam seen from the bottom, Electricité d'Emosson SA, M. Darbellay, 1969

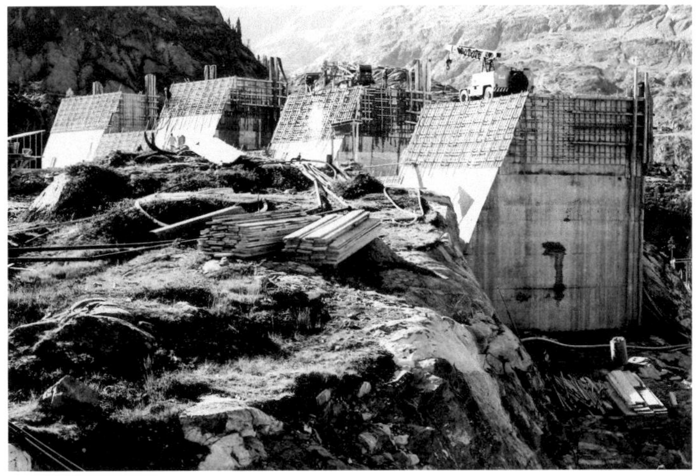

Emosson dam, first concrete right bank, Electricité d'Emosson SA, M. Darbellay, 1972

How can we understand the singularity of this networked landscape? From water retention by dams to more or less distant homes, this ensemble composed of a myriad of underground or aerial tentacles forms what the historian of technology Thomas P. Hughes calls a "large technical system."[3] The term designates both the physical network formed by infrastructures and various objects and the system of natural resources to which it is linked, or the conventions established by a whole group of human actors who have agreed to make it work (definition of the notion of power, economic model of investment or marketing, etc.). This is again an eloquent expression, because it reminds us that the networked landscape is not only a technical device but also something that requires us to deal as much with nature, its resources and resistances, as with other human beings to make such a delicate mode of functioning "hold."

As a networked landscape, a combination of geology and human intervention, the Emosson complex is in this respect an exemplary manifestation of these large technical systems which illustrate the way in which engineering has domesticated nature. Arched concrete vaults, interlaced cavities and tunnels, pumping stations, a series of pylons, and various cables form a curious framework that is progressively hybridized with the almost timeless mineral substrate of the Alps. This multiple apparatus is the marker of what the American geologist Peter Haff calls the "technosphere," a term that designates the physical part of the environment affected by man-made modifications: from agricultural land to transport and energy networks, including all sorts of machines of varying sizes.[4] As both the infrastructure of the daily life of the mountain populations and a playground for tourists, the Emosson complex and its evolutionary trajectory illustrates in its own way the way we try to inhabit the world, dealing with its opportunities and its limits. If the situation here seems stable and controlled, as Beatrice Gorelli and Keiichi Kitayama's project invites us to see, what will be its evolution due to the mutations in progress, between water scarcity and the environmental crisis? Will it progressively become a marker of an era of limitless resources? Or will it be a matter of rethinking its general structure? Let's bet on the point that the networked landscape of Emosson will continue to evolve, according to the constraints and human ingenuity regarding working with the elements. And that it will undeniably constitute a monument of choice for the hypothetical archaeologists of a distant future.

Nicolas Nova (HEAD – Genève, HES-SO)

水の巨塔、電動の大聖堂、景観的装置、水の番人、コンクリートモンスター。水力発電ダムを指す言葉は幅広く、豊富にある。そして、特にアルプスでは、高地の自然の中にダムが存在するため、その無機質な背景とダムの堂々とした景観が対照的だ。

ダムを表す言葉は、それを使う人の視点によってネガティブであったり強調的であったりと、巨大なダムの存在や外観に対して異なる意味合いを持つ。ダムの隠喩は、その土地における電力や水を貯めるという機能的な性質はもちろん、ダムの建築学的な佇まいにも関連づけられており、複合的な象徴的表現になっている。その点で、現代のおごりが具体的に表れている。過酷な環境の中にダムを建設し、その建設のために長い歴史の上に築かれたアルプスの地質に手を加え、また、スイス、ヴァレー地方出身の作家モーリス・シャッパ[1]が言うように、ダムが村の人々を動揺させてしまったのだ。ダムには直接的な有用性以上に、歴史的価値の側面があるのは疑いようもない。観光地、行楽地、思い出の場所という役割が次第に与えられるようになったのだ[2]。また同時に、自分たちの環境を凌駕する人類の力を称え、またおそらくはダム建設における尽力やそれに起因する悲劇を記憶に留めておこうという側面もある。

世界のネットワーク化を特徴とする文化から借用すると、こうした隠喩は非常にローカルなものであり、その土地のダムという建築物だけに注目しているということに気づかざるを得ない。電話回線によるネット接続に始まり、位置情報システムまであらゆるコミュニケーションシステムを知る21世紀の立ち合い人として、私はこうした隠喩表現がダムの所有権やネットワーク性を軽視しているように感じる。もちろん、20世紀にこうした巨大なダムが急ピッチで造られたことは、宗教的な大建造物や産業用機械など歴史的価値を見出されたこれまでの代物や、巨人やモンスターといった奇抜な伝説の生き物を引き合いに出すほどに衝撃が走ったというのも理解できる。しかし、私がエモッソンのようなアルプスのダムを見たときに印象に残ったのは、その本質であり、複雑な絡み合いの中に置かれた挿入物であるということだ。この織り合わせが、その土地においては建造物やそれらを結びつける空洞や配管、地域的には発電や電力による交通インフラ、もっと広い範囲で言えば領土内の多くの遠隔地への電力供給に繋がるといった具合に、さまざまな規模の地理に影響を及ぼす。ダムの風景は厳密にはローカルではなく、ネットワークの中に存在する。つまり、私たちはネットワーク化された風景を論じているのだ。

エモッソンは、その広大な空間を占める存在感ゆえに、そして間違いなく長い期間にわたって存在し続けているがゆえに、地質学上の新しい時代区分「人新世」(地球上の人類が将来的に重大な地質学的な変化をもたらすほどの強い影響を与えている時代を表す)の典型である自然と技術のハイブリッドという特別な象徴となっている。ここにベアトリーチェ・ゴレッリと北山景一によって記録された、この新しい姿こそが、この場所をユニークなものにしているのだ。

水力発電技術の画期的な一例であるエモッソンの基盤は、エモッソン、バルベリン、ヴュー・エモッソンの3つの主要なダムから成る複合体である。いずれも複数のトンネルと空洞で繋がり、揚水発電所ナント・ド・ドランスを地下に設けることで複合的なダムの強度を補っている。その姿は地上では分からないが、地下で見ると壮大だ。

1 Maurice Chappaz, *Chant de la Grande Dixence* (Lausanne: Payot, 1965; new edition by Éditions de L'Aire, 1995).

2 Tristan Loloum, "La vie touristique des grands barrages hydroélectriques," *Mondes du Tourisme* (2016), 12.

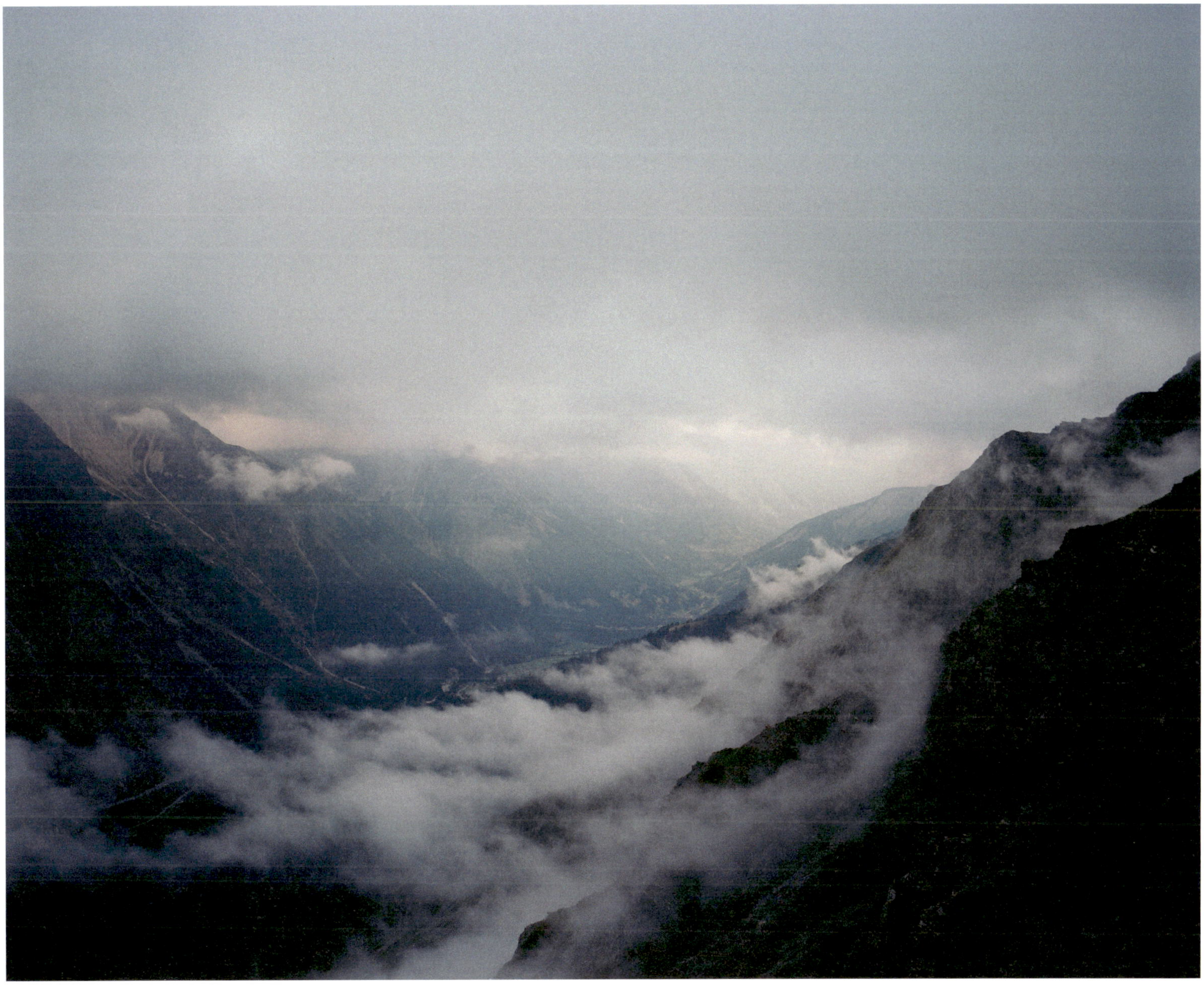

建設後に何度かリニューアルされた3つのダムから、さまざまな要素を集結して大規模な貯水と水力発電を統合したシステムが作られた。地元であるフランスとスイスのネットワークという形のこのシステムは、近隣のトリヤン、ロニャン、ル・トゥール、アージャンティエールの氷河から、運河と導水路を使って水をエモッソン・ダムに取り込んでいるという事実によっても示されている。約40kmに及ぶ複数の水路をエモッソン・ダムに引くというこの巨大ネットワークによって、176km^{2}のエリアから水を集め、エモッソンの貯水容量2億2500万m^{3}を満たしているのだ。

そうしたローカルネットワークは、表面しか見ない者にはほとんど見えない。ダムの荘厳な存在感を示すアーチならまだしも、これほど複雑な仕組みを一見しただけで把握するというのは難しい。取水装置は目立たず、トンネルや水路は山中に埋められている。また、3つのダムのうち、バルベリン・ダムは、エモッソン・ダムの貯水レベルが非常に低くなった時にしか見ることができない。谷に水を放出する作業中は、エモッソン・ダムが稼働している違った姿を目の当たりにできるが、放出が終了すると元の静かな山奥に戻る。

この複合的なダムは観光スポットとして、基礎構造を見学でき、その歴史的価値としての側面にも触れることができる。バルベリン・ダムの建設に必要な資材や人員を運ぶために作られた古いケーブルカーは、観光客が利用できるようになっている。ダムの上まで行くことができるミニケーブルカーも同様である。パノラマビューを堪能できるトロッコ電車では森やトンネルを抜けて風を受けながら、山腹ルートからのモンブラン山塊の眺望を楽しめる。

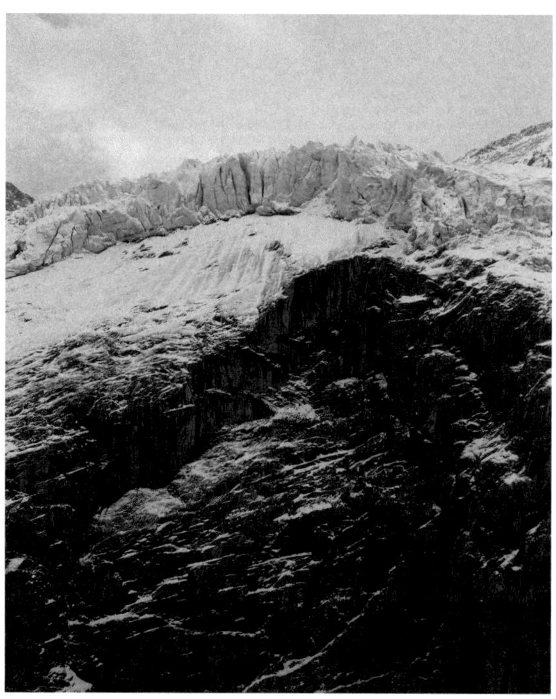

もし水力発電と交通の複合施設が明白なローカルシステムを形成するなら、広大なネットワークへと拡大していくことを忘れてはならない。まず、オート・サヴォワ（フランス）のヴァロルシーヌと、マルティニ（スイス）のラ・バティアスにある2つの隣り合う発電所で電気を作ることができるように、近くに送水管を設けた。こうしたインフラが、例えば、ラ・バティアスでは非常に高電圧の地中送電線を引き、地上の見える場所には鉄塔とケーブルを組み合わせた従来の架空送電線を引くことで、双方を送電ネットワークに繋ぎ、住民、公共機関、一般企業に電力を供給できるようさまざまな送電所へと繋ぐ。日常で使うあれこれをコンセントに差し込めば、山奥の風景の中にあるネットワークを強く意識することもなく繋がるのだ。

ネットワーク化された風景の特異性について、どうすれば実感できるか。水を貯めたダムから多かれ少なかれ遠方の家庭まで、この集合体は地下と空中に無数の触手を伸ばし、技術史学者のトーマス・P・ヒューズが言うところの「大規模な技術システム」を形成している。この「大規模な技術システム」[3]とは、インフラやさまざまな要素によって形成される物理的なネットワークとそれに繋がる天然資源のシステム、もしくは、それを機能させることに同意する当事者のグループ全体によって決められた約束事（権利の概念、投資やマーケティングの経済モデルなどの定義）の双方を指している。これもまた、説得力のある表現である。ネットワーク化した風景は技術的な側面だけでなく、人類が「自然を守る」ことをデリケートに捉えているように、自然やその資源、その抵抗力と向き合う必要があることを思い出させてくれるからだ。

3 Thomas P. Hughes, "The Evolution of Large Technological Systems," in *The Social Construction of Technological Systems*, *eds.* E. Bijker, Th. P. Hughes, and T. Pinch, (Cambridge, Mass.: MIT Press, 1987).

4 Peter Haff, "Technology as a geological phenomenon: Implications for human well-being," *Geological Society, London, Special Publications,* 395 (2014), 301–309.

エモッソンの複合的なダムは、ネットワーク化された風景であり、地質に人間が介入したものとして、その点においては、技術が自然を手なずける方法を示す、こうした大規模な技術システムの模範を明示している。アーチ状のコンクリート天井、入り組んだ空洞やトンネル、ポンプ室、連なった鉄塔、多様な電線と、ほぼ永久とされるアルプスのミネラル基質を革新的に掛け合わせることで興味深いフレームワークを形成している。こうした複合的なシステムは、アメリカの地質学者、ピーター・ハフが作った「テクノスフィア（技術圏）」という概念の指標となる。テクノスフィアとは、人工的に手を加えることで環境の物理的な側面に影響を与えることを指し、農地から交通やエネルギーネットワークまで、さまざまな規模のあらゆる種類の機械化が含まれる[4]。山に住む人々の日常生活を支えるインフラと観光客の行楽地として、エモッソンの複合的なダムとその進化の軌跡は、地球の可能性と限界に向き合いながら、私たちが地球に生きようとする術を独自の方法で示している。ベアトリーチェ・ゴレッリと北山景一のプロジェクトが我々に示すとおり、エモッソンの状況が安定かつ統制されているように見えるのであれば、水不足と環境危機が交錯する中で起こっている異変によって、その進化はどのようになるのか。次第に、エモッソンは資源に限りがない時代の指標となるのか、はたまた、エモッソンの構造全体を見直すことになるのか。ネットワーク化されたエモッソンの風景が、限りある資源とその保護に取り組む人間の知恵によって発展を遂げることに賭けてみようではないか。そして、それが歴史的価値のあるものと判断するのは、紛れもなく遠い未来の考古学者たちなのである。

ニコラス・ノヴァ　ジュネーヴ造形芸術大学 HEAD、西スイス応用科学芸術大学 HES-SO

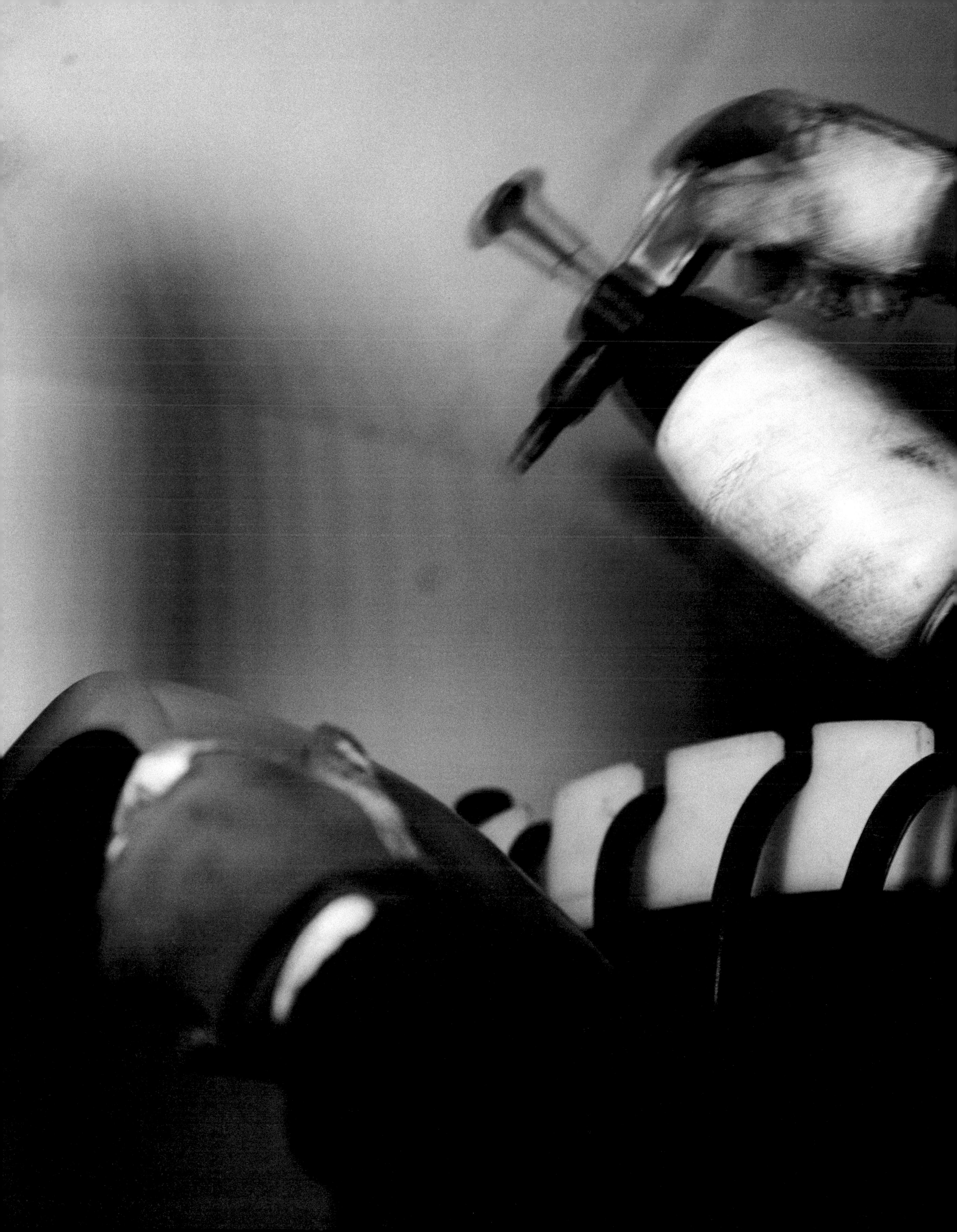

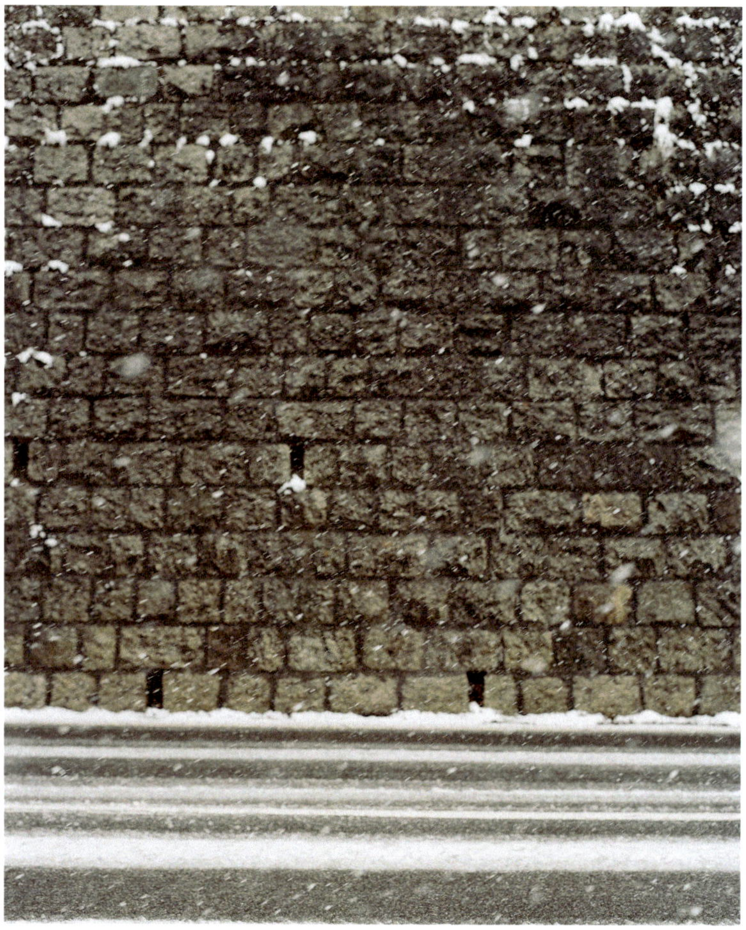
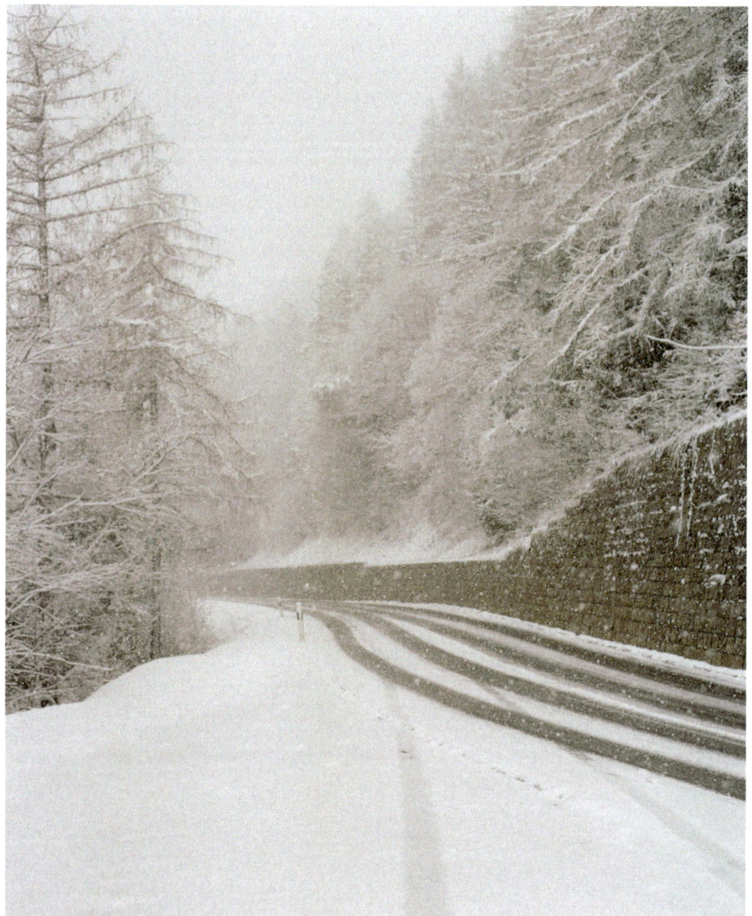

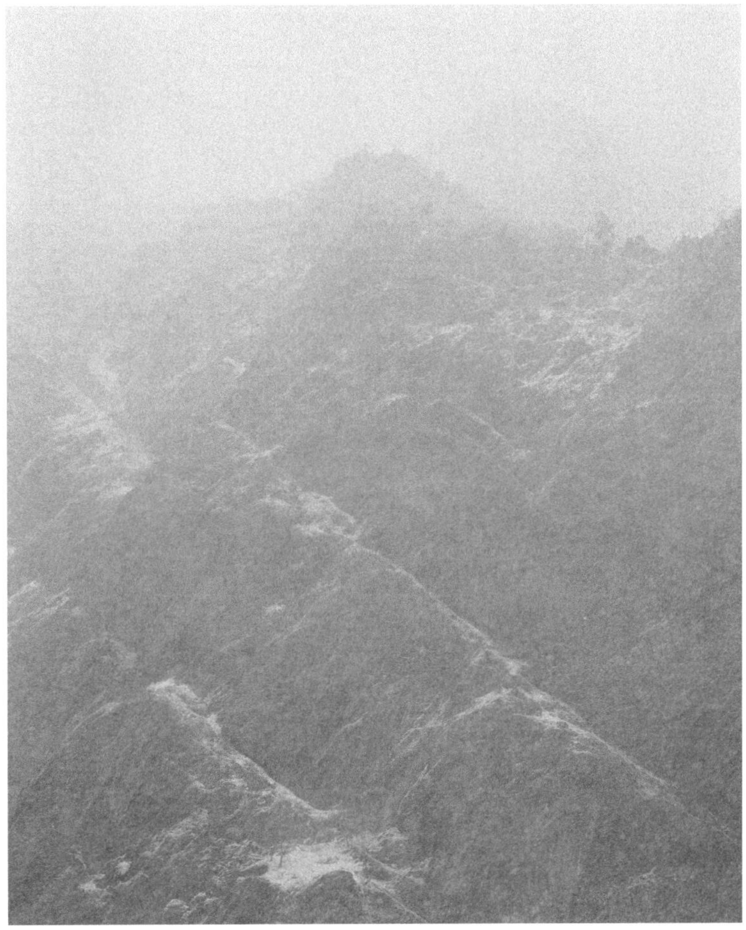

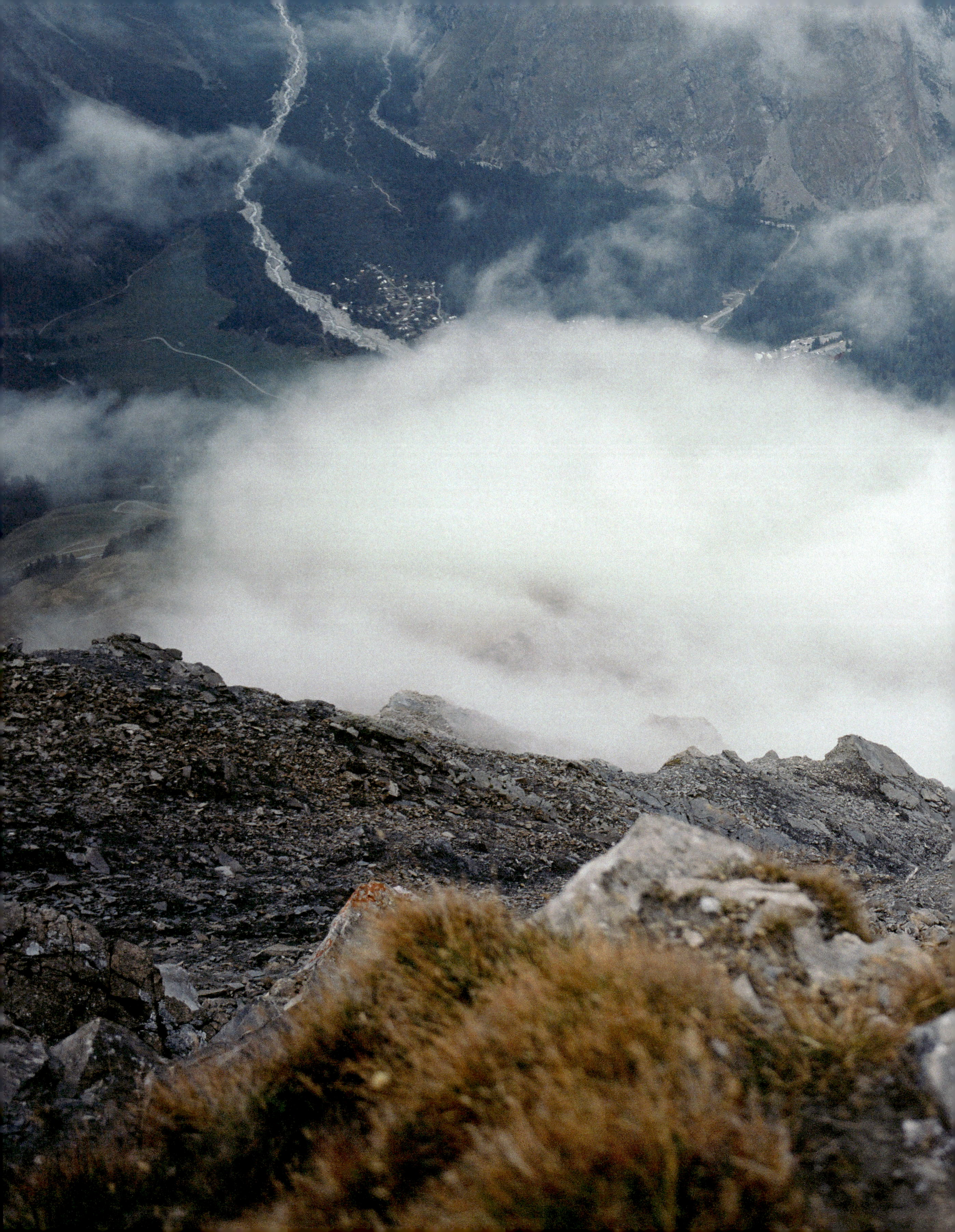

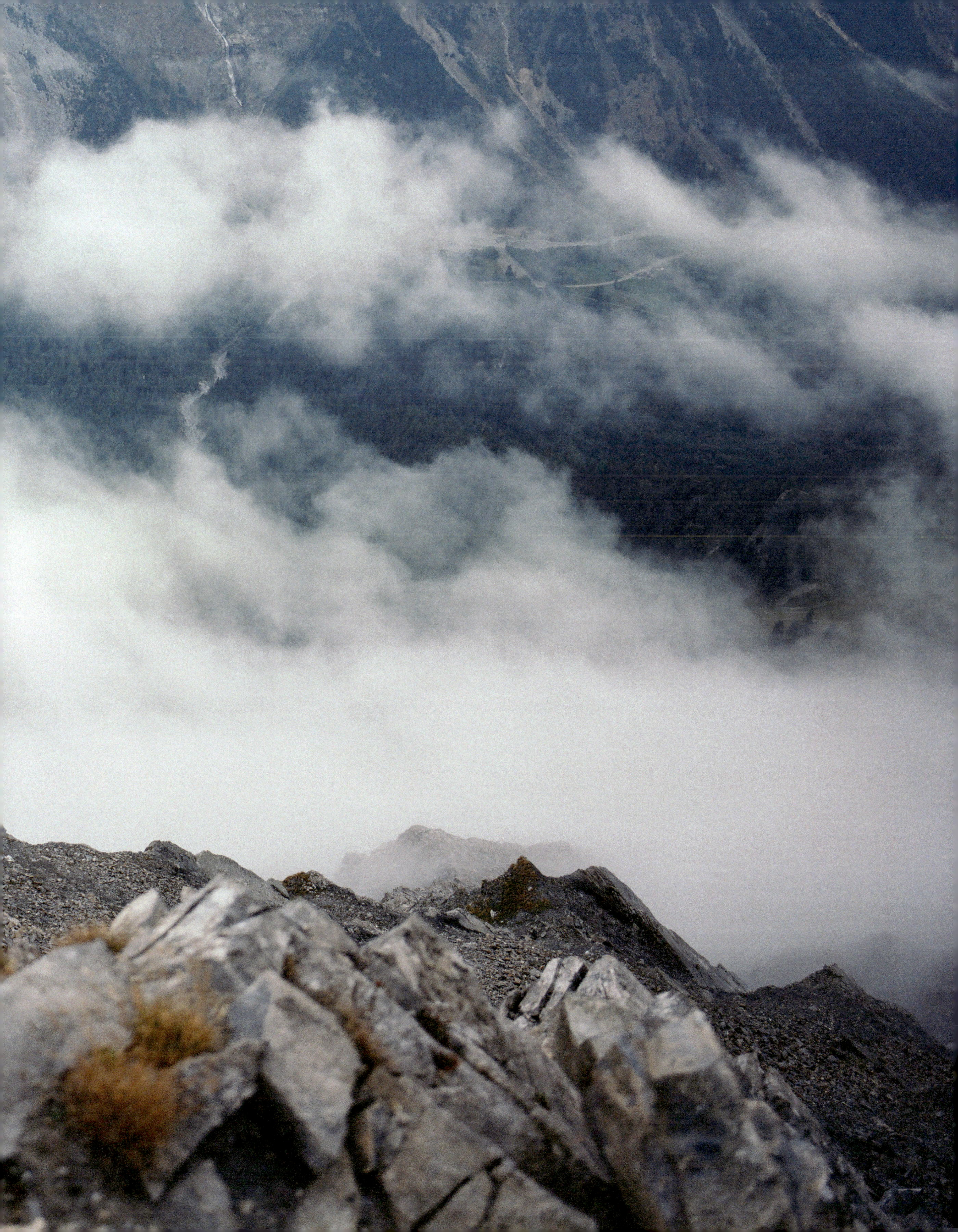

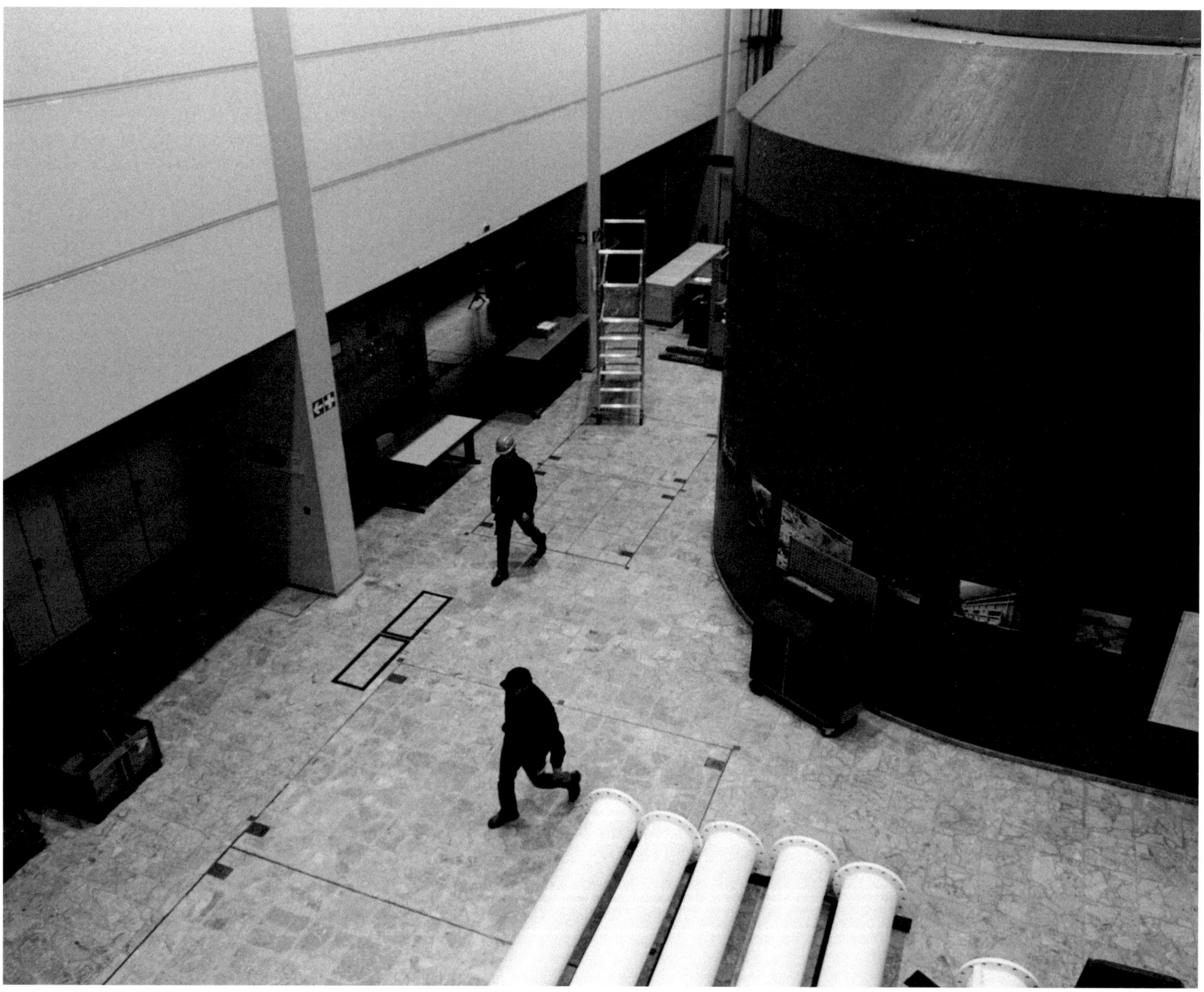

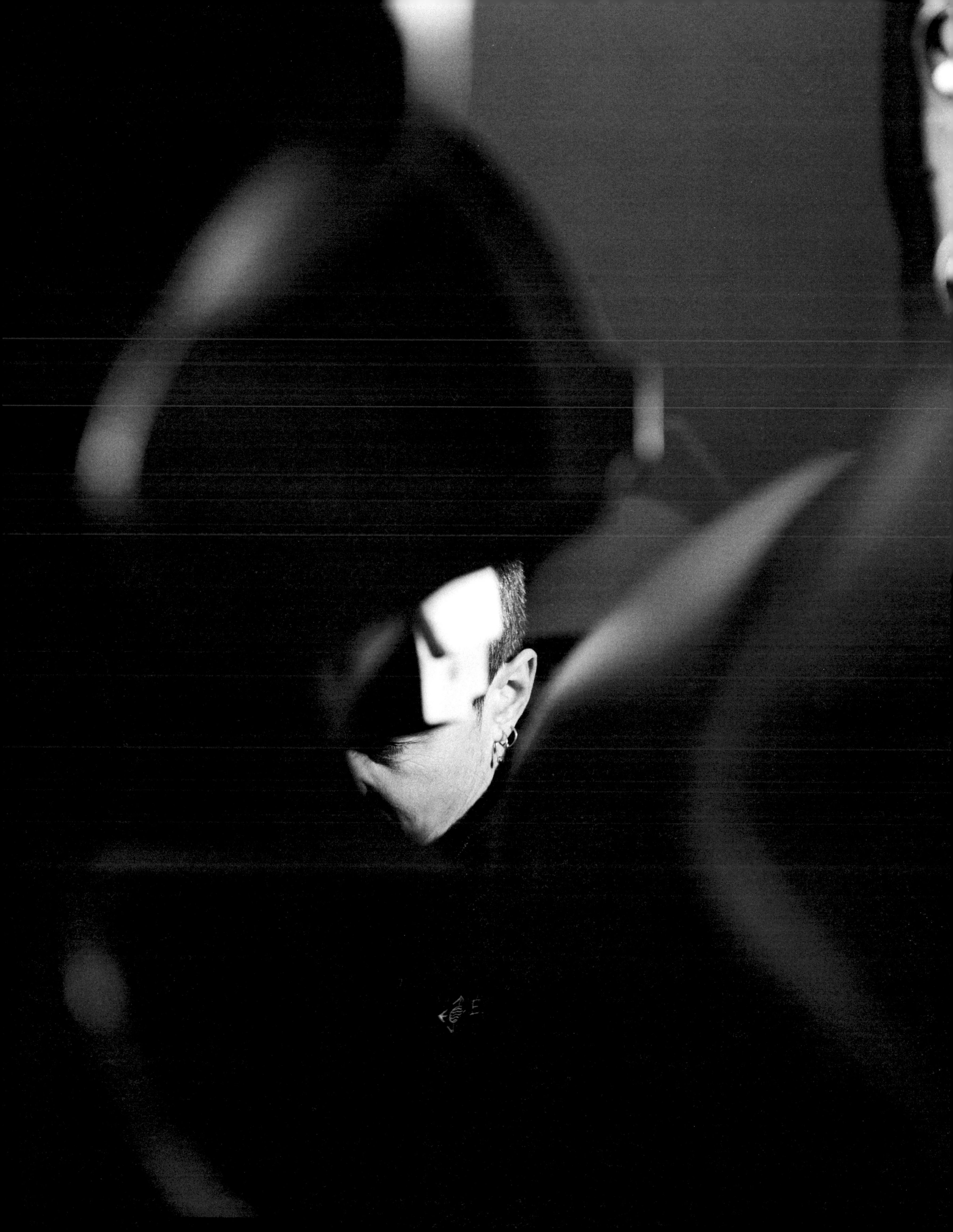

The Earth, the Sea, the Spheres
Speak of your power.

PIETRO METASTASIO

From the deep shadows of the silent fir grove
I tremblingly behold you, summit of eternity,
Dazzling peak, from whose punishing height
My spirit soars into the infinite.

Who sunk the pillar deep into the Earth's lap,
Which, for millennia, firmly supports your mass?
Who raised high in the vault of the ether
Mighty and bold your radiant countenance?

Who poured you down from the realm of eternal winter,
O torrents of peaks, with thunderous clamor?
And who commands loudly with the voice of omnipotence:
"Here shall rest the frozen waves!"

Who traces the path for the morning star there,
Who adorns with blossoms the hem of eternal frost?

To whom resounds in terrifying harmonies,
Wild Arveiron, your tumultuous waves?

Jehovah! Jehovah! it cracks in the bursting ice;
Avalanches thunder as they roll down the chasm;
Jehovah! it rustles in the bright treetops,

Whispers in trickling silver streams.

Friederike Brun, (*03.06.1765, † 25.05.1835)
"Chamonix at Sunrise, In May 1791," in *Gedichte*, ed. Friedrich Matthisson (Zurich: Orell, Gessner, Füssli and Co., 1795)

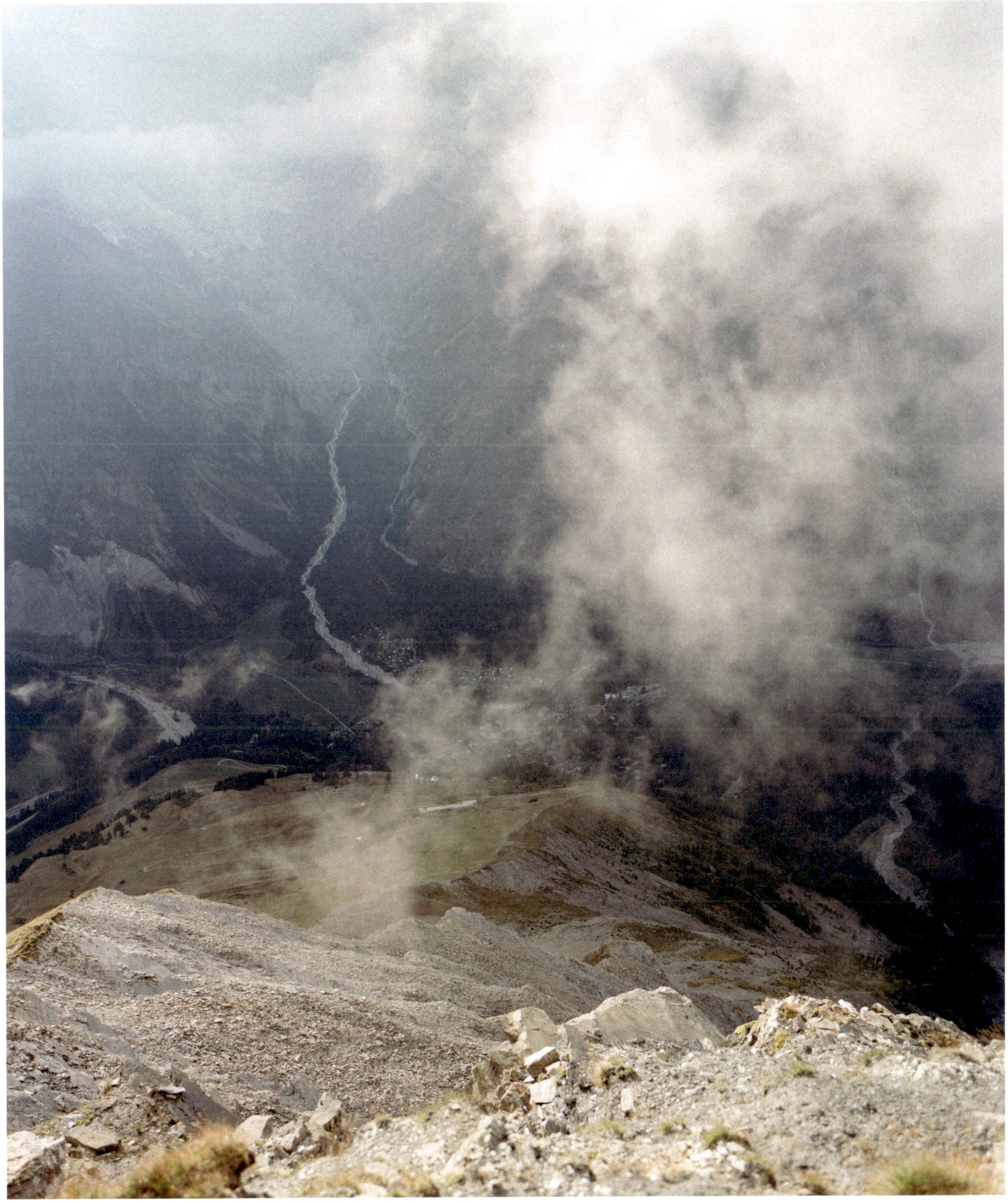

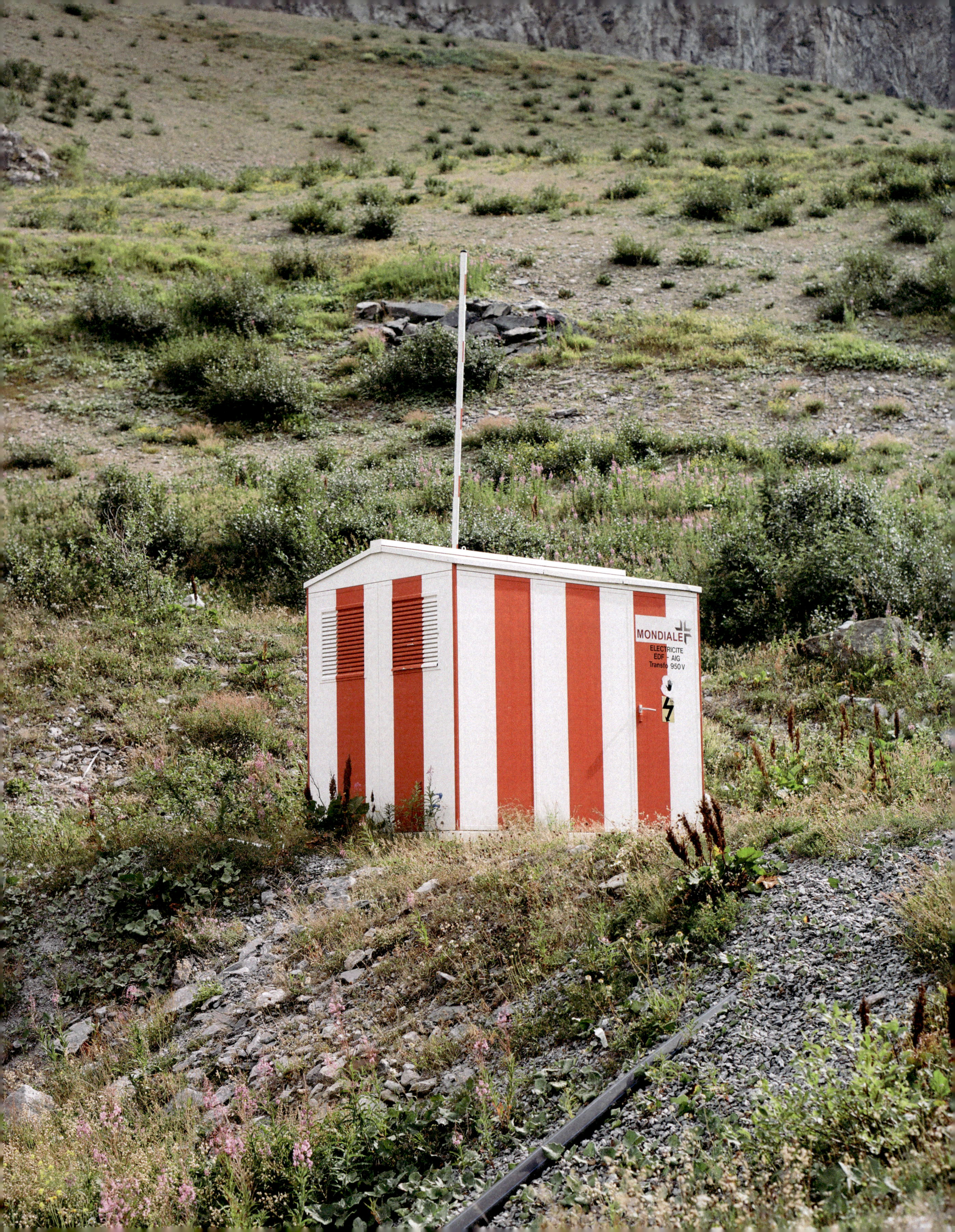
MONDIALE
ELECTRICITE
EDF - AIG
Transfo 950 V

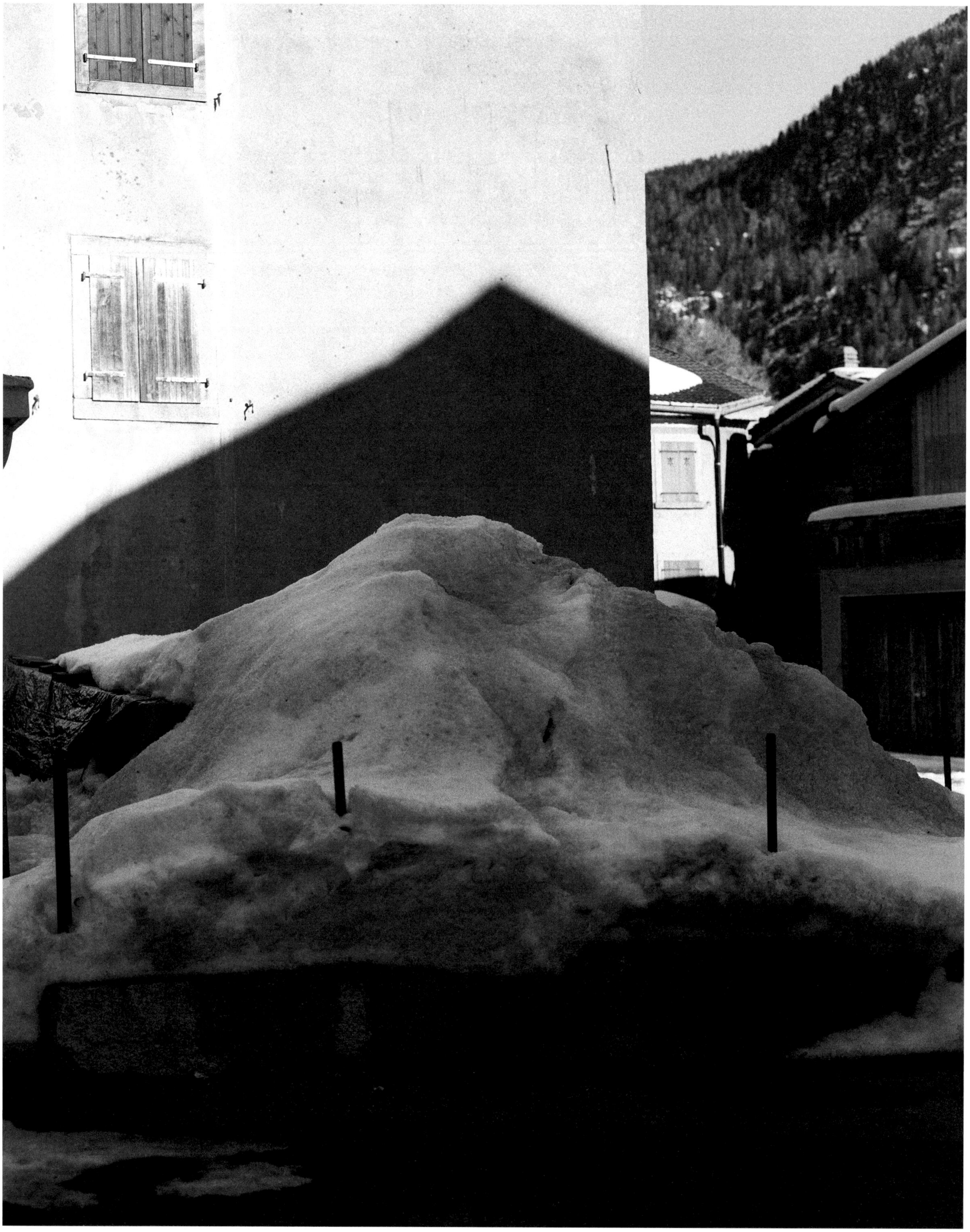

シャモニーの朝日
1791年5月

フリーデリケ・ソフィー・クリスティアーヌ・ブラン

「大地、海、天体は
人間の力を物語っている」メタスタジオ

静かなるモミ林の暗い影から
おののきながら君を見つめる
無限に続く頂　まぶしさを放つ山
その容赦ない高みから
この魂が際限なく舞い上がる

誰が大地の谷間に柱を深く沈めたのか
数千年の間　その塊を堅固に支えるものなのか
誰が天空という円天井の高みに昇らせたのか
強大で力強いその燦然とした容貌なのか

誰が果てしない冬世界から君を注ぎ込んだのか
ああ轟きながら山々からほとばしる水だろうか
そして誰が全能の声で勇ましくも指揮したのか
「いざ　この凍てつく波を鎮めよう」と

誰が明けの明星に道を示すのか
永遠なる厳寒の麓を花で飾る者だろうか
恐ろしいほどの協和音で誰に向かって鳴り響くのか
激しいアルヴェロン川よ　その荒々しい波なのか

全能の神ヤハウェよ！　砕けそうな氷に亀裂が入る
深い裂け目を転がり落ちる雪崩が轟音を立てる
ヤハウェよ！　まぶしい梢の間でサラサラと音がする
銀色の川が流れ落ちながら囁いている

Friederike Brun, (*03.06.1765, † 25.05.1835)
"Chamonix at Sunrise, In May 1791," in *Gedichte*, ed. Friedrich Matthisson (Zurich: Orell, Gessner, Füssli and Co., 1795)

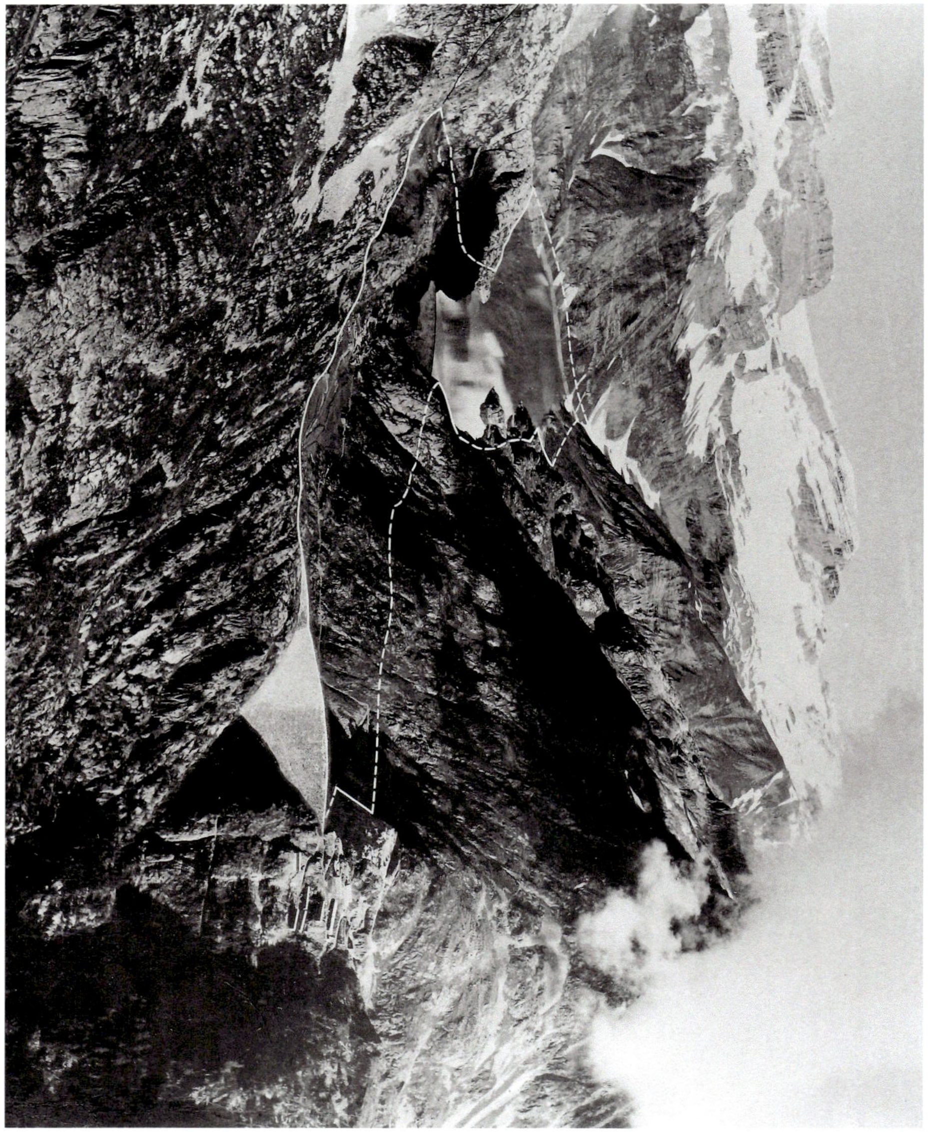

Emosson dam before being filled with water,
Barberine dam in the background,
Electricité d'Emosson SA,
M. Darbellay, 1968

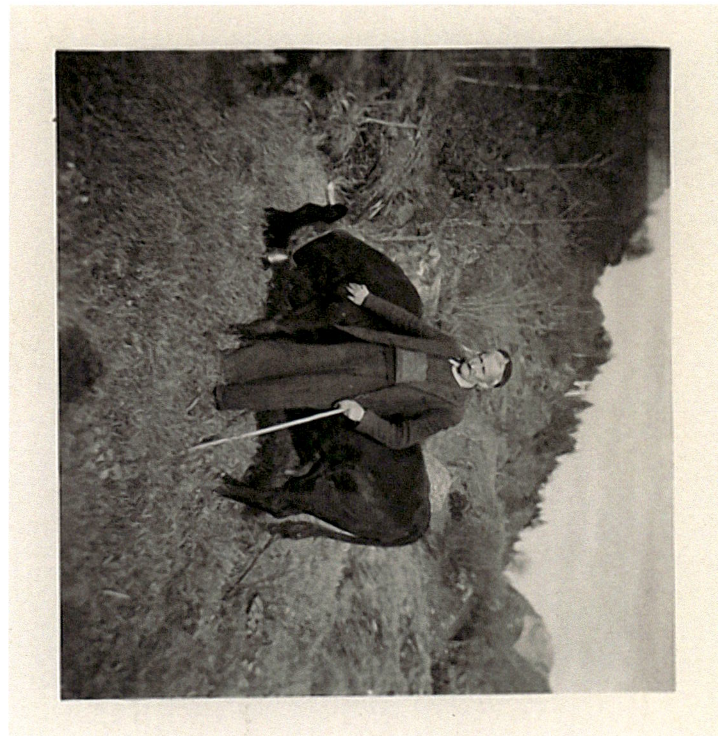

Armand Gay des Combes,
Bruno's grandfather, Finhaut, 1951

Bruno's grandfather returned from France to Switzerland in 1928 with one goal in mind: to find work at the Barberine dam. Upon arrival, however, he discovered that they were no longer hiring workers, leaving him without prospects. Undeterred, he decided to explore other possibilities in the region and eventually settled at the foot of the valley, in the charming village of Finhaut. There, he made a life for himself as a farmer and also worked as a porter in the hospitality industry. He worked tirelessly to provide for his family. Although he did not achieve his initial goal, his perseverance paid off in the long run. His son participated in the construction of the Emosson dam, which became an essential hydroelectric power station in the region. Eventually, Bruno himself found employment at Emosson SA, creating a legacy that spans several generations. The Barberine dam still stands today, reminding us of the paths our lives can take.

It is a powerful example of how small decisions can shape the course of our lives, and how hard work and determination can lead to unforeseen opportunities. Bruno's grandfather's journey is an inspiring testament to these truths, and his family's legacy continues to be felt in the region to this day.

ブルーノさんの祖父は、バルベリン・ダムに携わる仕事を見つけようと一念発起して、1928年にフランスからスイスへ戻った。ところが帰国すると、すでに作業員を雇っていないことが判明し、将来の見通しが立たなくなる。それでも彼はひるまず、この地域で他の可能性を模索することにした。そして最終的に、魅力的なファンオーの村にある谷の麓に落ち着いた。そこで農場主として生計を立てながら、ホテル業界でドアマンの仕事もして、家族を養うために休むことなく働いた。当初の目的を果たすことはできなかったものの、その不屈の精神は、長年かけて報われることになる。彼の息子が、スイスの重要な水力発電所となるエモッソン・ダムの建設に関わったのだ。そして今、ブルーノさん自身もエモッソンSA社の社員となり、世代を超えて伝わる思いを受け継いでいる。バルベリン・ダムは現在もあり、私たちの人生の道しるべを想起させる。

小さな決断が人生の行方を左右すること、そして懸命に働き、強い意志を持つことが、思いもよらないチャンスをもたらすことを教えてくれる印象的な話だ。ブルーノさんの祖父の人生は、こうした真実を体現した感動的な証しであり、彼の家族が受け継ぐ思いは、今もこの地で生き続けている。

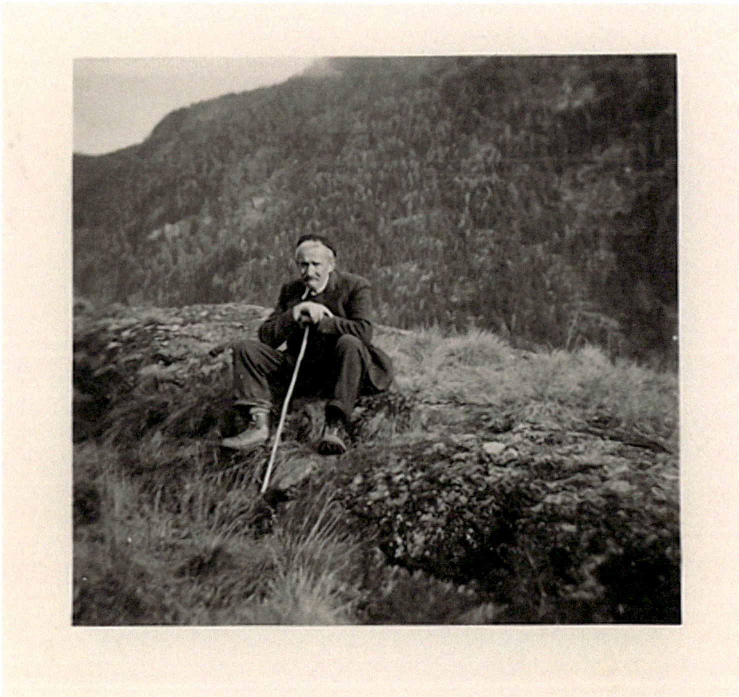

Armand Gay des Combes,
Bruno's grandfather, Finhaut, 1951

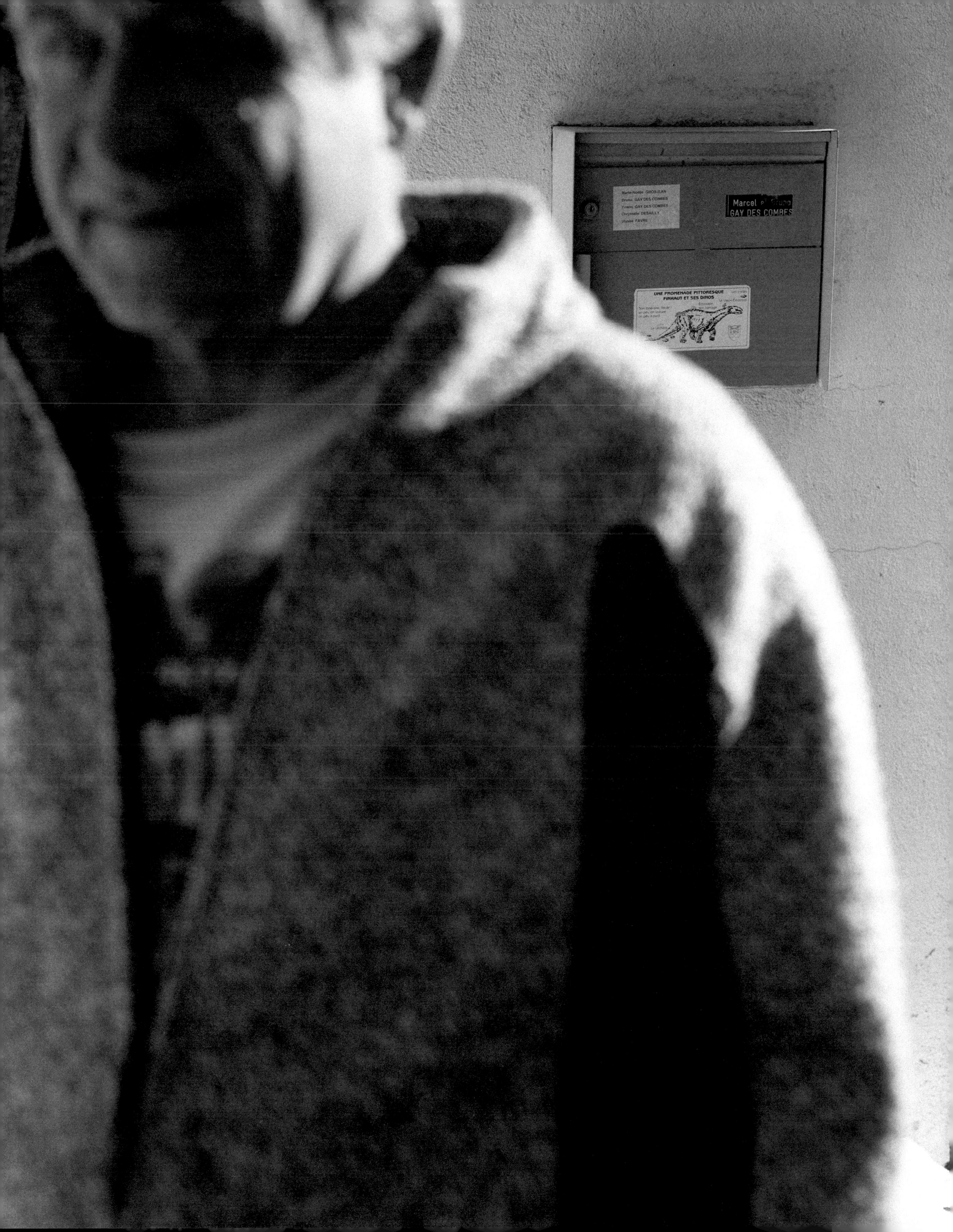
Marie-Noëlle GROSJEAN
Bruno GAY DES COMBES
Yoann GAY DES COMBES
Chrystelle DESAILLY
Ulysse FAVRE
Marcel
GAY DES COMBES
UNE PROMENADE PITTORESQUE
FINHAUT ET SES DINOS

Armand Gay des Combe's
passport, Bruno's grandfather

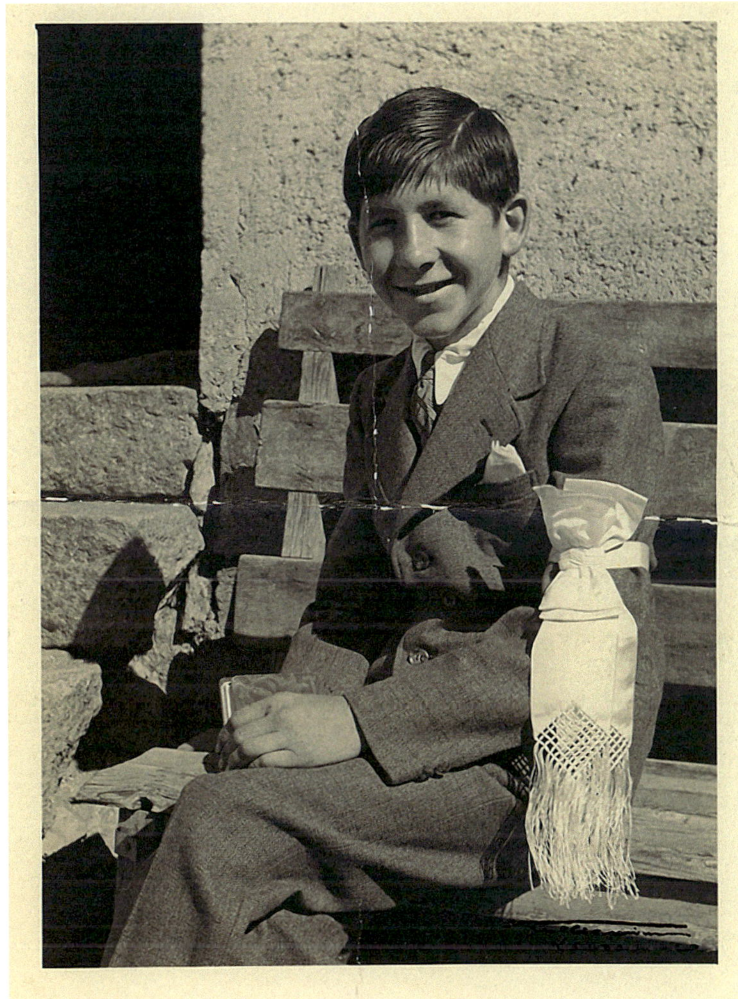

Marcel Gay des Combes,
Bruno's father, Finhaut, 1942

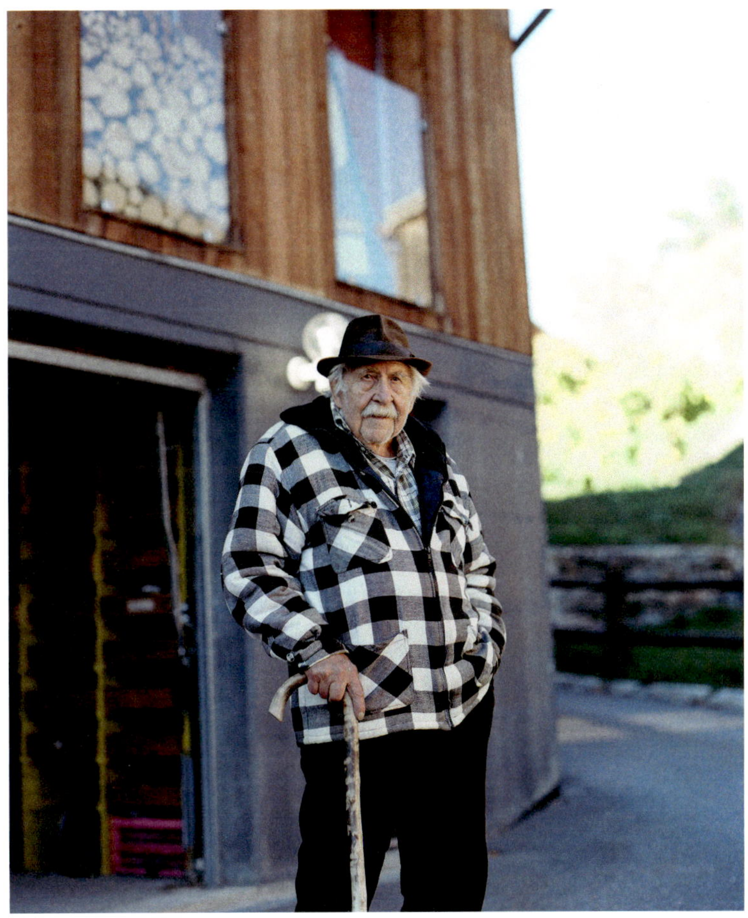

Marcel Gay des Combes,
Bruno's father, Finhaut, 2021

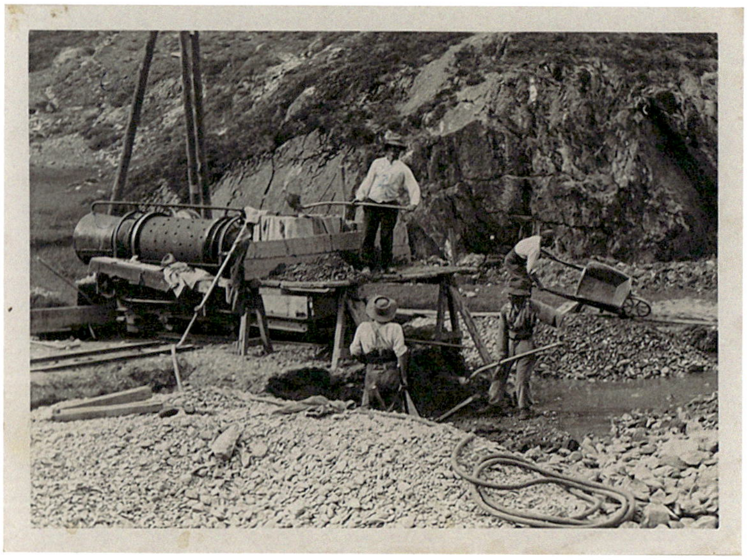

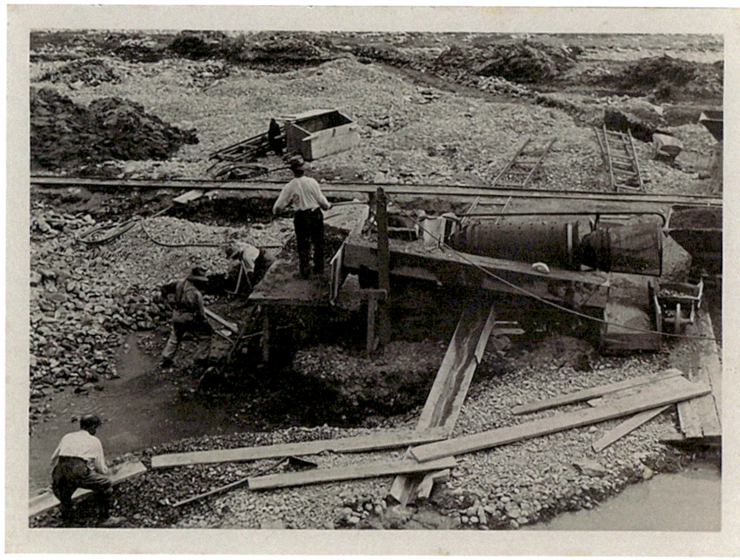

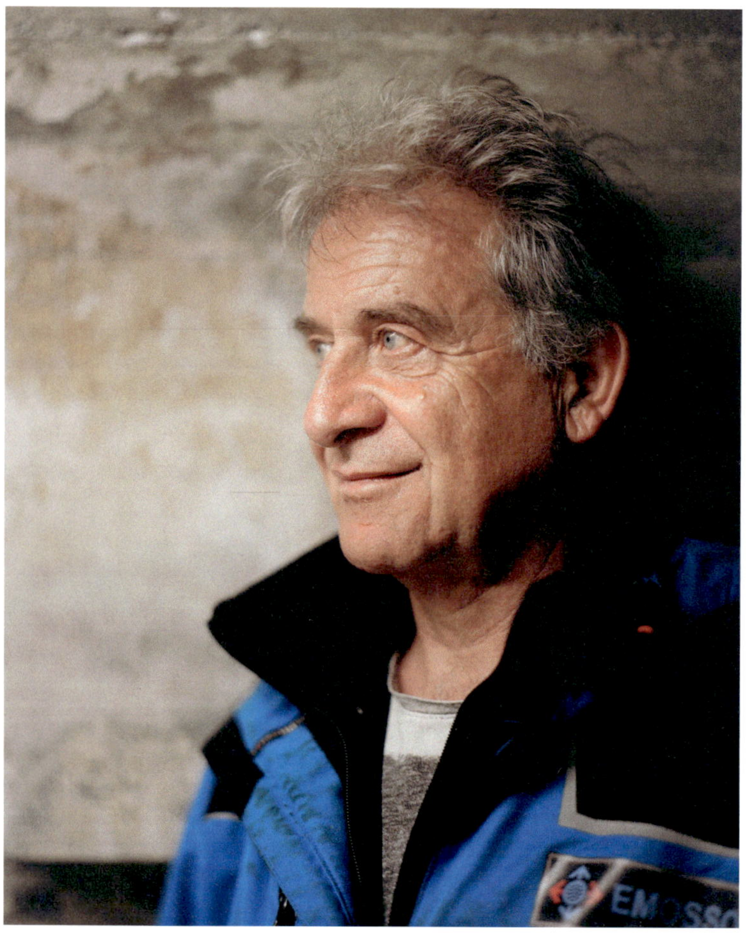
EMOSSO

Workers, Barberine dam
foundations, 1920

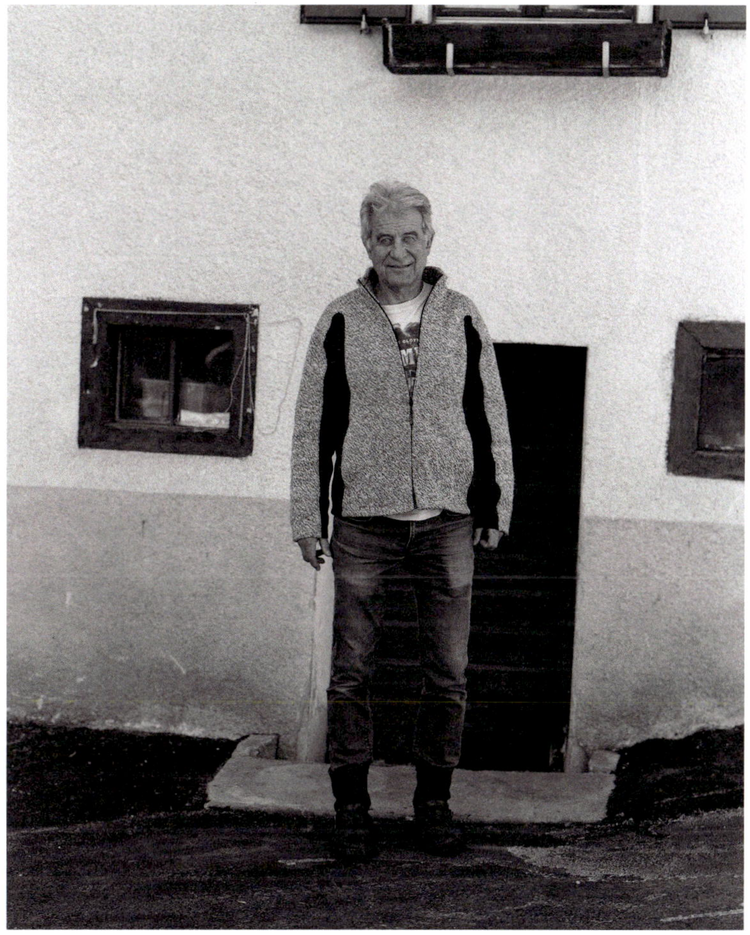

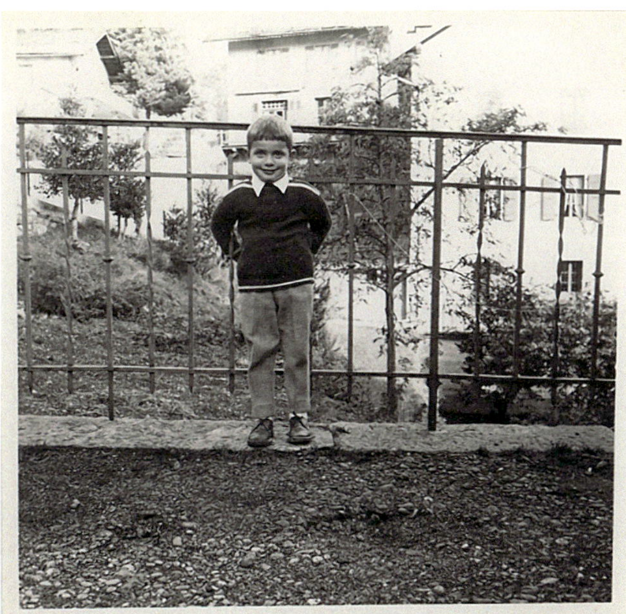

Bruno Gay des Combes,
Finhaut, 1963

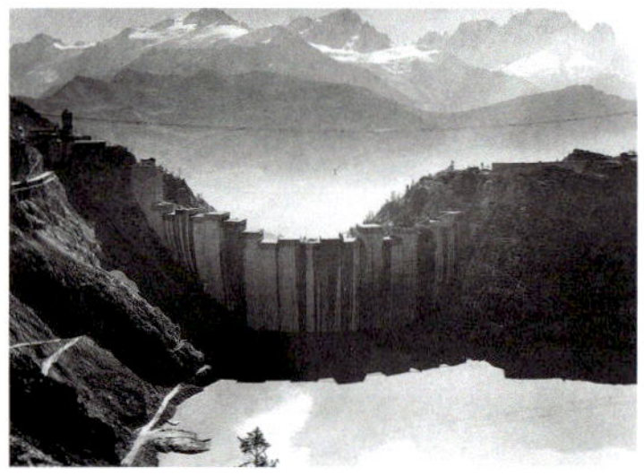

Emosson dam, view from the mountain, Electricité d'Emosson SA, M. Darbellay, 1972

BK What drew you to this profession?

BGC It was chance! I had taken a sabbatical year, and left the city for the mountains, but then I needed to find work because I ran out of money.
I was very interested in mountain activities and wanted to come back to work in Finhaut, where I was born. At that time, there were very few opportunities as a mechanical draftsman in Valais, and it was thanks to the retirement of an Emosson employee who was in charge of shift service, stopping and starting the machines for electricity production, that I was able to join the company. Other tasks were added to my work, such as creating plans for the construction of power plants and the dam, as well as monitoring and measuring the dam.

BK How many years have you worked for Electricité d'Emosson SA?

BGC I started working at Emosson on December 1, 1981, and retired in June 2022, so I worked for Emosson for forty-one years.

BK What resonates with you and pleases you about dams?

BGC The diversity of work at Emosson. I initially started working in shifts at the plant: night, morning, or afternoon—depending on the work, with weekly rotations. Later on, I also took on the tasks of measuring the dam and monitoring charts that tracked the movements of the dam wall. Over time, with experience and modernization of work tools, I handled computer work and statistics, analyzing water inflows and electricity production. Then, for twenty years, I resumed the measurement and control work of the dam. I love the dam because it's beautiful and located in a superb place at the beginning of the valley, offering a panoramic view of Mont Blanc. The dam itself is like a giant holding back water; I find it impressive, and it's always a pleasure for me to come and take measurements in this natural environment, two days a month.

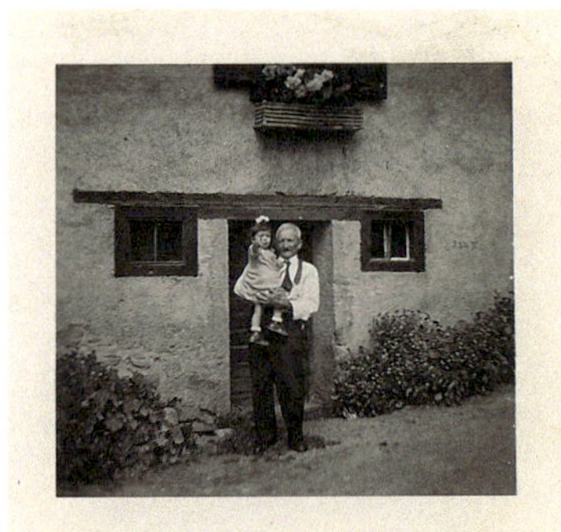

Armand Gay des Combes with Bruno's sister, Finhaut, 1953

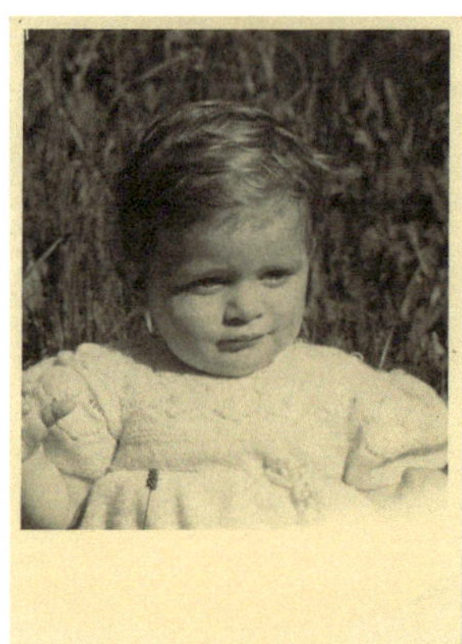

Bruno Gay des Combes, Salvan, August 1958

BK Does your entire family live in Finhaut, and why did they settle there?

BGC My whole family is from Finhaut; it's our village. My family name, Gay des Combes, is also from Finhaut. My mother was Italian, and my father learned the cooking profession in Neuchâtel. He worked in hotels and later in restaurant cars. For twenty years, my parents, my sister, and I lived in Lausanne, where I attended school and completed my apprenticeship. Then I wanted to come back to Finhaut, so I found work at Emosson, and have been there ever since. My father, who was about to retire, also returned to Finhaut.

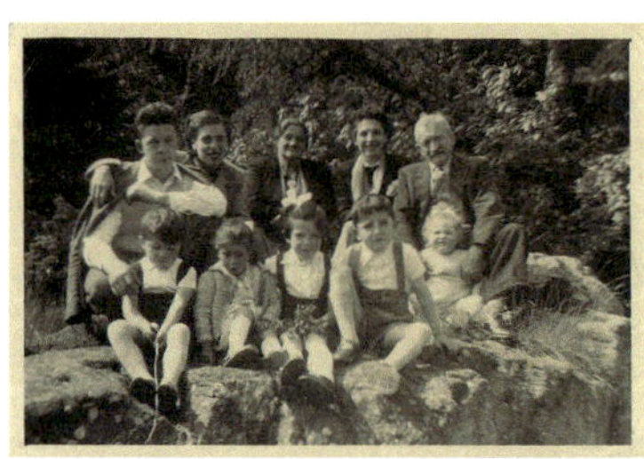

Bruno's family, Finhaut, August 1953

BK Now that you are retired, what is your hobby?
BGC I'm just beginning my retirement, so it's a transitional phase, and for now, I'm still doing the same work. But I enjoy doing things, being outdoors, skiing, hiking, or cycling.

BK What were the most challenging moments for you in this company?
BGC There were more or less small events, but no major event that particularly marked me. When we were on shift duty, we worked alone, and sometimes it felt unusual to be alone at night in the plant. When there were mechanical problems, it generated stress. We had to call colleagues to repair the machines and manage the basins that could overflow, but usually we managed to get the machines running again and handled electricity production with the dispatchers. Sometimes, when there were problems, I would think that I had earned my monthly salary in just one night. But over the years and with experience, you learn to better manage unforeseen circumstances.

BK Have you noticed any changes in the work environment related to climate change? For example, does the dam's water evaporate quickly, or do glaciers melt too fast?
BGC You don't see the evaporation of the lake water. I was in charge of hydraulic statistics, and for Emosson, the main source is glacier water, so as long as the glaciers are there, there will be enough water. However, in the future, this quantity could decrease. Currently, the dam is full, but it could be only half full depending on needs and electricity production. Overall, at Emosson, we cannot yet measure the extent of these changes.
BK What do you think of hydroelectric power and other means of energy production?
BGC An installation like Emosson and hydraulic power is like a wildcard in a deck of cards. If we compare a car and a truck, the car can accelerate very quickly, like the hydraulic power of a dam that can be quickly available when needed, and the truck is more comparable to nuclear power. We need all these different ways of producing electricity; hydropower alone is not the solution and is not sufficient. Therefore, we need both environments to make it work, as well as all other energy sources like solar or wind power.
BK How do you feel when you are inside the dam? Are you aware of any safety issues?
BGC The dam is in very good condition; it's an asset to be maintained. The delicate phase is the initial filling, and then for fifty to sixty years, there shouldn't be any problems. After that, over the years, issues may arise, but this dam is still very young.

BK What do you think of people who oppose the dam, such as environmentalists and nature advocates?
BGC Like everything else, it's a matter of choice. If you need comfort to live, you have to accept that something will be degraded in return. For example, my grandfather had a small stable with cows located on the Emosson alpine pasture, and with the construction of the dam, the pasture was flooded, and he lost his stable. But in return, we were able to have electricity and a bit of comfort. As a side note, my grandfather worked in the hospitality industry as a doorman in Finhaut during the summer and in southern France during the winter, where he met my grandmother. After the First World War, life was not easy in France. In 1921, his brother told him about the construction of the Barberine dam and offered him a job there. Unfortunately, when he arrived in Switzerland, the dam was already completed, and workers were being hired to monitor the dam and operate the power plants. In the end, he couldn't find work so he kept a few cows and lived as a farmer, but he continued to work in hotels in Finhaut during the summer.

BK: Beatrice Gorelli and Keiichi Kitayama
BGC: Bruno Gay des Combes

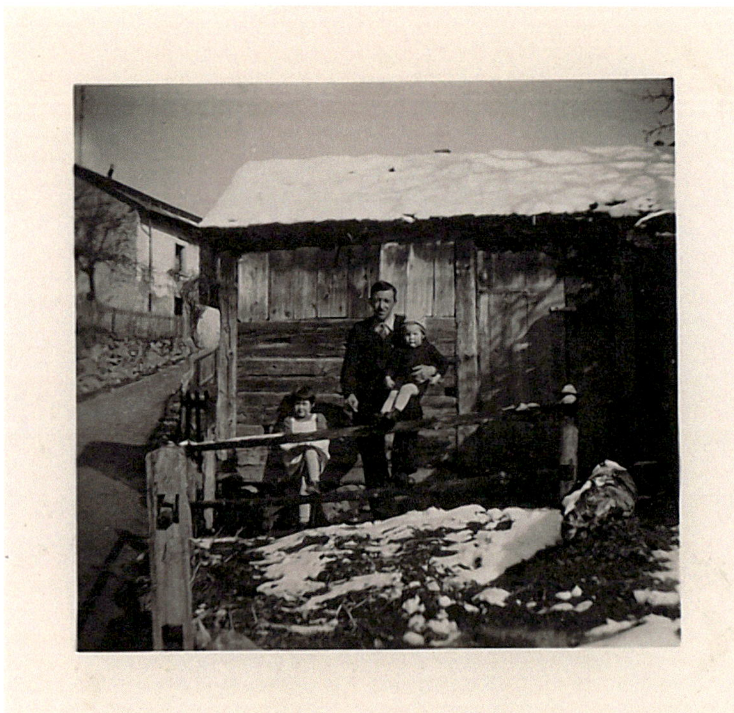

Jean Gay des Combes's,
Bruno's uncle's chicken coop,
Finhaut, 1951

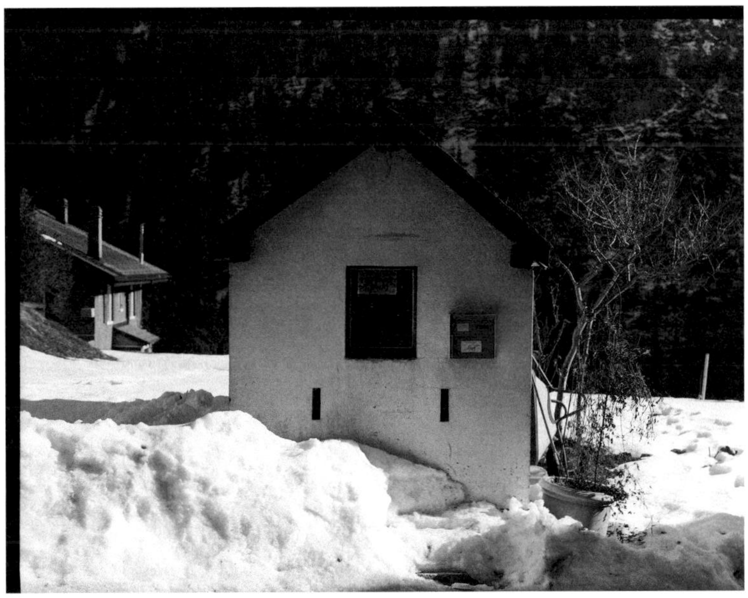

Jean Gay des Combes's
chicken coop, now a storage area,
Finhaut, 2023

BK　この仕事に就いたきっかけは?

BGC　チャンスに恵まれていましたね!1年間の有給休暇を取って街を離れ、山地へ行ったんですが、お金が尽きてきたので働かなくてはいけなくなって。山での活動に非常に興味を持っていたし、故郷のファンオーに戻って働きたいと思っていたのですが、当時、ヴァレー地方では機械製図士の仕事はほとんどありませんでした。発電の変換業務や機械操作に携わっていた1人のエモッソン社員が退職したおかげで、この会社に入社することができたのです。私は、発電所やダム建設の計画の作成、ダムの監視や測定といった他の業務も行っていました。

BK　エモッソンSA社での勤続年数は?

BGC　1981年12月1日に入社して2022年6月に退職したので、41年、働いたことになります。

BK　ダムで共感したり、喜びを感じたことは?

BGC　エモッソンの業務は多岐にわたります。最初は発電所で変換の作業をしていました。1週間ごとに、夜、朝、昼とローテーションしながら行う業務でした。その後、ダムの測定やダム壁の水位動向の監視も行いました。やがて経験を積み、作業ツールが近代化されるのに伴って、コンピューター作業や統計を担当し、水の流入量や発電量を分析するようになりました。それから20年間は再び、ダムの測定と管理業務に就きました。私はダムが大好きなんです。美しいですし、谷の上という最高の場所にあって、モンブランのパノラマビューを楽しめますから。ダムは水を湛える巨人のようで、感動します。1か月に2日、ダムを訪れ、自然の中で計測に携わることは常に喜びでした。

BK　ご家族は皆さん、ファンオーに住んでいるのですか?ここに落ち着かれた理由は?

BGC　私の家族は全員、ファンオー出身で、この村が故郷です。私の姓である「Gay des Combes」も、ファンオーに由来しています。母はイタリア人で、父はヌーシャテルでプロの料理人として修業していました。ホテルで働いたあとは、キッチンカーで働いていましたね。両親、姉、私は20年間ローザンヌで生活していました。私はそこで学校へ通い、実習期間を終えました。その後、私はファンオーに戻りたいと思い、エモッソンで職を見つけて現在に至ります。定年退職を間近に控えた父も、ファンオーに戻ってきました。

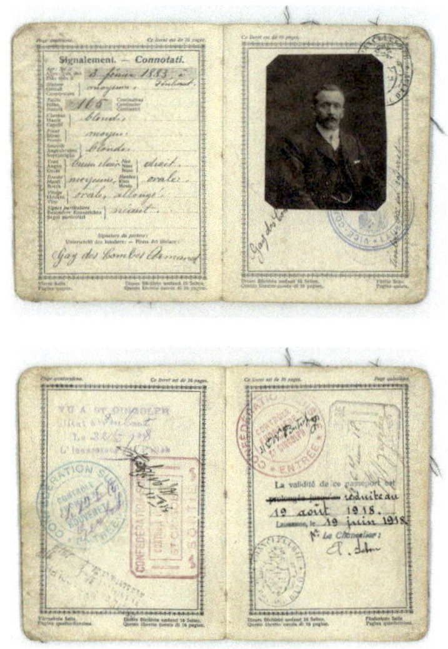

Armand Gay des Combes's passport

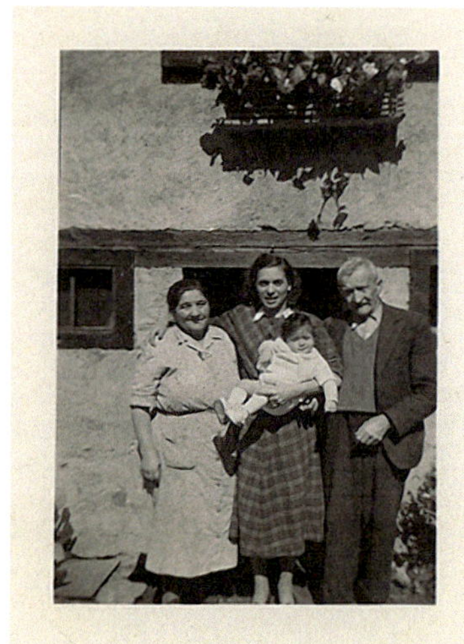

Bruno's grandparents, mother, and sister Finhaut, 1953

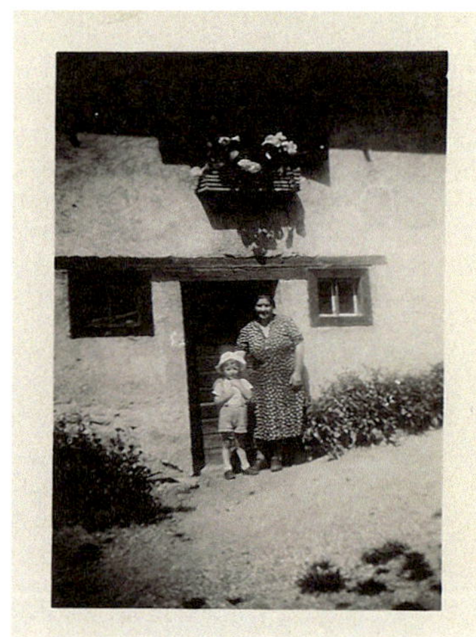

Bruno's cousin and grandmother, Finhaut, 1953

BK 今は退職されていますが、趣味はありますか?
BGC 退職したばかりなので、移行段階といったところです。今のところはまだ、同じようなことばかりしていますが、アウトドアやスキー、登山、サイクリングを楽しんでいます。

BK この会社で、最も大変だったことは?
BGC 多かれ少なかれ些細な出来事はありましたが、特に記憶に残るような大事件はありません。変換業務は独りでの作業だったので、夜中に発電所に独りぼっちという状況を異様に感じたことはありましたね。機械トラブルが発生した時は緊張しました。機械を修理するのに同僚に電話して、決壊の恐れがあるダムを管理しなければいけません。でも大抵は、機械を再稼働して、給電担当者と一緒に発電処理ができましたけどね。問題が発生したときは、一晩で1か月分の給料を稼いだと思うこともありました。ですが、何年も経験を重ねると、予期せぬ事態でもうまく切り抜けることを学ぶものです。

BK 気候変動に伴い、ダムの水の蒸発や氷河の融解が加速したなど、職場環境に何らかの変化を感じたことは?
BGC 湖の水の蒸発は確認されていません。私が担当していたのは水圧の統計データです。エモッソン・ダムの場合、主要な水源は氷河の水なので、氷河がある限り水は十分にあります。しかし、将来的にその水の量が減る可能性もあります。現在、ダムは満水ですが、水の需要と発電量によっては半分だけになってしまうかもしれません。概してエモッソンでは、まだそうした変化の影響について測定できていません。

BK 水力発電と他の発電システムについて、どう思いますか?
BGC エモッソン・ダムのような設備と水力発電は、カードでいうと万能なワイルドカードのようなものです。車とトラックに例えるなら、必要に応じて速やかに供給できるダムの水力発電は、俊敏に加速できる車であり、原子力発電はトラックに近いと言えます。私たちには、あらゆる発電手段が必要なのです。水力発電だけでは解決できませんし、十分ではありません。ですから、どちらの発電所も稼働させる必要があり、太陽光や風力など他のエネルギー源も必要です。

BK ダムの内部にいるときはどんな気持ちですか?安全性について問題はありませんか?
BGC ダムの状態は、非常に良好です。保全された資産ですから。初期の貯水は慎重さが求められる時期ですが、その後50〜60年間は、問題は起こりません。そこから何年も経過すれば、問題が起こる可能性はありますが、このダムはまだ非常に若いです。

BK 環境や自然保護など、ダムに反対する人たちについてどう思いますか?
BGC 何にでも言えることですが、何を選ぶかということですね。快適な生活が必要なら、その見返りとして何かが損なわれることを受け入れなければいけません。例えば、私の祖父はエモッソンの高山牧草地に牛と小さな家畜小屋を持っていました。しかし、ダムの建設で、牧草地は水中に沈み、家畜小屋を失いました。でもその見返りに、電気とささやかな快適さを手に入れることができたのです。余談ですが、祖父は、夏はファンオー、冬は南フランスで、ドアパーソンとしてホテル業界で働いていました。そこで、祖母と出会ったのです。第一次世界大戦後、フランスでの生活は苦しいものでした。1921年に祖父の兄からバルベリン・ダムの建設の仕事を勧められたのですが、残念なことに祖父がスイスに着いたときにはダムはすでに完成していて、ダムの監視と発電所の運営に携わる人しか雇っていませんでした。結局、祖父はダムの仕事に就くことができず、数頭の牛を飼って農場主として生計を立てました。それでも、夏の間はファンオーでホテル勤務を続けていたのです。

BK: Beatrice Gorelli and Keiichi Kitayama
BGC: Bruno Gay des Combes

EMOSSON AU SEUIL DE LA NOUVELLE ANNÉE

par R. Wetter, ingénieur EPFZ

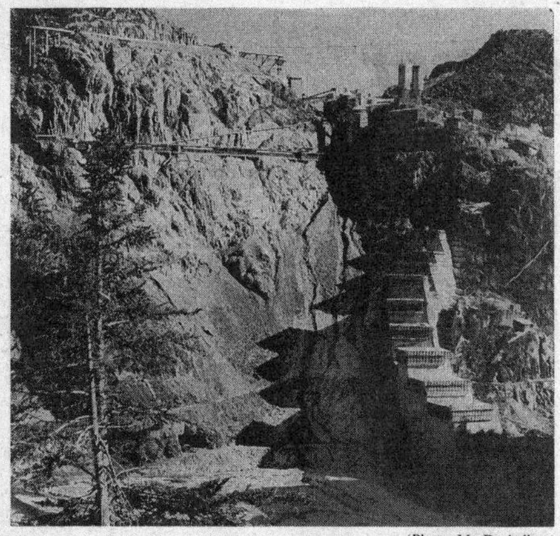

(Photo M. Darbellay.)

Rive gauche du barrage d'Emosson avec installations de chantier ; état hiver 1970-1971.

Le travail sur la plupart des chantiers d'Emosson, arrêté pour les fêtes de fin d'année, va reprendre sous peu, marquant plus ou moins le début de la deuxième mi-temps de la période de construction proprement dite de cet aménagement hydro-électrique à accumulation de haute chute dans le Bas-Valais et les Alpes savoyardes. Une bonne raison pour faire le point de l'état actuel de la réalisation.

L'état d'avancement des travaux à fin 1970 est le suivant :

Au chantier du barrage, 210.000 m3 ont été mis en place, principalement sur la rive gauche, tandis que sur la rive droite les travaux d'excavation pour les fondations du barrage se poursuivent.

Pour le collecteur est, l'avancement dans la galerie principale atteint pour tous les chantiers environ 14,3 km. sur une longueur totale de 17,7 km. Les tronçons La Fouly-Saleina et Trient-Les Esserts sont perforés et le bétonnage du revêtement est en cours, ainsi que la construction du barrage de la prise de La Fouly.

Au chantier du bassin de compensation des Esserts de la chute secondaire Les Esserts-Le Châtelard, les travaux d'excavation et le remblaiement des digues sont terminés. Le puits incliné des Esserts aboutissant à la Centrale de Châtelard-Vallorcine a été percé le 22 décembre 1970.

Pour la chute Emosson-Le Châtelard, la chambre des vannes est terminée et le pont roulant installé. Le blindage du puits incliné conduisant à la centrale de Châtelard-Vallorcine a été terminé le 16 décembre 1970.

Au chantier du bassin de compensation de la chute Le Châtelard-La Bâtiaz, l'excavation est pratiquement terminée et le mur de rive en cours d'exécution. La perforation de la galerie d'amenée atteint environ 7 km. sur une longueur de 10 km. au total. Les travaux de revêtement entre le Châtelard et Tête Noire sont en cours. Les conditions géologiques rencontrées et des venues d'eau importantes ont nécessité un léger changement du tracé de la galerie. Au puits incliné de Ravoire, le bétonnage du radier ainsi que les travaux de soutènement sont pratiquement achevés. Les fondations de la centrale de La Bâtiaz sont terminées et le bâtiment en éléments préfabriqués est en montage.

En France, la prise d'eau sous-glaciaire sur la dérivation permanente d'Argentière est achevée. Les ouvrages souterrains d'accès et d'adduction des eaux du Tour sont en cours d'achèvement, de même que ceux placés en tête aval du collecteur sud (cheminée d'équilibre, siphon, chambre des vannes, etc.).

Sur l'ouvrage proprement dit du collecteur sud, l'avancement atteint environ 6,8 km. sur une longueur de 8,2 km. au total. La zone particulièrement délicate du synclinal de Chamonix a été franchie sans difficultés notables, comme ce fut également le cas au collecteur est. Le puits incliné de Belle-Place, d'une longueur de 1,83 km. a été perforé le 15 novembre 1970, et le blindage mis en place à la partie inférieure de cet ouvrage atteint une longueur de 700 m. Au collecteur ouest, le tronçon aval en direction du bassin d'accumulation d'Emosson a été perforé à fin août 1970 ; il reste encore un peu moins de 2 km. à perforer en direction du Vallon de Bérard.

A la centrale de Châtelard-Vallorcine, les bâtiments de la salle des machines, de la chambre des robinets [illegible] et couverts. Les ponts sont tous opérationnels et le montage des groupes a débuté en novembre 1970.

Dans l'ensemble, les travaux progressent d'une façon satisfaisante, bien que des imprévus d'ordre géologique aient conduit à adapter le programme sur deux chantiers et font prévoir un retard de quelques mois dans l'achèvement de certains travaux.

Tandis que pour les questions techniques, Motor-Columbus, Ingénieurs-Conseils S.A., et EdF, REH Alpes Nord, trouvent des solutions raisonnables, la situation sur le marché de la main-d'œuvre n'est pas sans préoccuper le maître de l'œuvre, bien que ces problèmes soient ceux des entreprises. Les mouvements contre la surpopulation étrangère et les mesures prises par le Conseil fédéral dans ce domaine, qui vont de pair avec l'amélioration des conditions de travail dans les pays fournissant cette main-d'œuvre, ont des répercussions souvent difficiles à surmonter, principalement dans les chantiers de haute montagne. Il faut espérer que les autorités responsables sauront prendre à temps les mesures et la libéralité qui s'imposent.

Rappelons peut-être que l'accumulation considérable d'Emosson de 225 millions de m3 d'eau correspondant à 683 millions de kilowatt-heures sera exploitée en commun par les Chemins de fer fédéraux et Electricité d'Emosson S.A., société de droit suisse, dont le siège social est à Martigny. Electricité d'Emosson S.A. est une société composée de trois partenaires : Aar et Tessin, S.A. d'Electricité S.A. (ATEL) Olten (25 0/0), Motor Columbus, S.A. d'Entreprises Electriques, Baden (25 0/0), ainsi que d'Electricité de France, Paris, la grande société nationale assurant la majeure partie de la production, du transport et de la distribution de l'énergie électrique en France (50 0/0).

Les partenaires, en tant qu'actionnaires de la société, se sont engagés à prendre livraison de la totalité de l'énergie produite, au prorata de leur participation au capital-actions, ainsi qu'à payer dans la même proportion les charges annuelles. La société bénéficie de deux concessions de forces hydrauliques, une fédérale et une française, ainsi que d'une convention franco-suisse réglant toutes les questions d'utilisation des forces hydrauliques nécessaires à la réalisation de l'aménagement provenant à parts égales des deux pays et déclarant l'aménagement d'utilité publique. En effet, cet aménagement produira une qualité d'énergie très recherchée. Elle peut non seulement être concentrée presqu'entièrement sur les mois d'hiver, mais également sur les heures journalières de forte consommation.

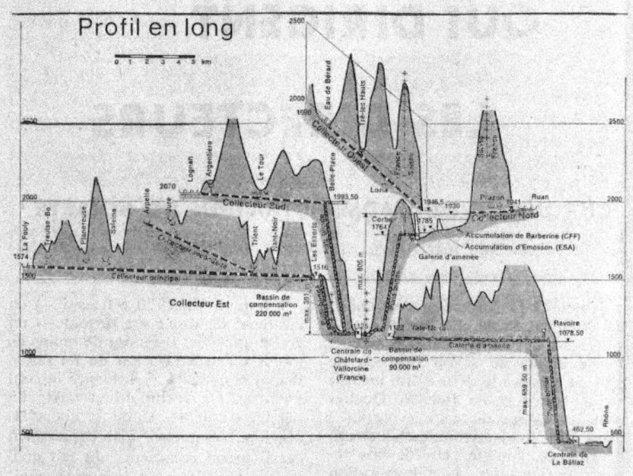

L'exploitation est également très souple, les groupes hydrauliques pouvant très rapidement s'adapter à de fortes variations de charge et démarrer en trois minutes environ. Un tel aménagement, dit de pointe, constitue ainsi un complément idéal aux centrales nucléaires futures dont les caractéristiques sont très différentes, mais qui sont la seule source d'énergie capable de reprendre la relève de l'équipement hydraulique dont les possibilités économiques sont pratiquement épuisées en Suisse. Il s'agit de construire tous les 12 à 15 ans des moyens de production de l'importance de toutes celles qui existent déjà, tout en conservant bien entendu ces dernières, et ceci aussi longtemps que le taux de croissance actuel se maintient.

La consommation par habitant d'autres pays, tels que la Norvège, le Canada, la Suède et les Etats-Unis [illegible] d'une part, et la propreté et la commodité de cet agent énergétique d'autre part, ne laissent pas entrevoir une saturation prochaine, au contraire. Comme les nouvelles sources d'énergie électrique ne sont pas meilleur marché que les anciennes, les investissements nécessaires croîtront très rapidement.

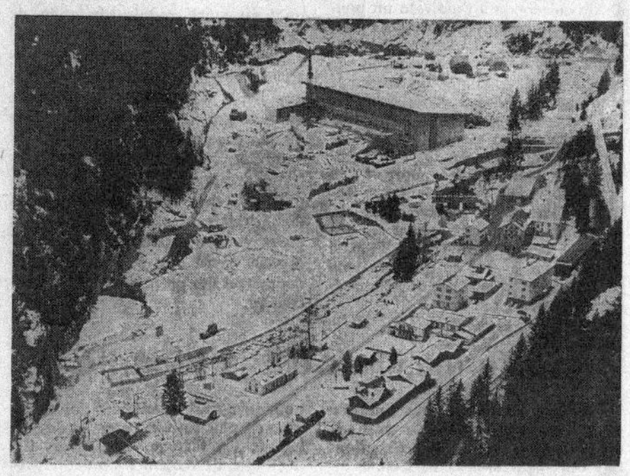

(Photo M. Darbellay.)

Centrale de Châtelard-Vallorcine (F) et bassin de compensation de Châtelard (CH) ; état hiver 1970-1971.

En ce qui concerne le financement d'Emosson, là également la première mi-temps est achevée. Le capital social, actuellement de 100 millions de francs, est libéré à 70 0/0, tandis que sept emprunts par obligations ont été placés sur le marché suisse, soit pour un montant total de 210 millions de francs. Le taux nominal moyen de ces emprunts est de 5,73 0/0.

La situation sur le marché suisse des capitaux, très fluctuant à la fin de l'année 1969, s'est stabilisée à partir du mois d'août 1970 à un niveau jamais atteint pour les emprunts de sociétés d'électricité. Cette hausse n'a rien de particulièrement inquiétant pour Emosson. étant donné qu'elle est générale, mais les répercussions sur le prix de l'énergie électrique seront inévitables et d'autant plus fortes que cette situation se prolongera.

La rentabilité de l'aménagement d'Emosson, calculée avec un intérêt déjà très élevé au départ, ne pourra que s'améliorer par rapport aux autres aménagements. Actuellement déjà, plusieurs gros aménagements construits dans les années 1950 à 1960 doivent procéder à la conversion de leurs emprunts à un taux qui passe souvent du simple au double. Il est bon de rappeler ici que la part des intérêts servis représente pour la plupart d'entre eux plus de la moitié du prix de revient de l'énergie.

Et pourtant, le financement par le marché des capitaux est la seule solution pouvant offrir à l'économie entière des prix d'énergie électrique avantageux. L'autofinancement complet conduirait à des majorations de prix très sensibles avec des répercussions incalculables pendant une longue période de transition. Il faut espérer que l'économie électrique pourra compter sur le marché des capitaux également à l'avenir.

En ce qui concerne les emprunts d'aménagements hydro-électriques, il est évident qu'ils restent tout aussi sûrs, même à l'avènement des centrales nucléaires. L'économie électrique ne se passera pas de centrales hydrauliques pour des raisons d'exploitation et ne les abandonnera jamais, puisque les charges de ces centralse sont pratiquement indépendantes que ces dernières fonctionnent ou non. Les emprunts de sociétés d'électricité sont de ce fait des titres de placement recherchés par les fonds de prévoyance et les titres d'Electricité d'Emosson S.A. ont été reconnus par le bureau fédéral des assurances à Berne comme valeur pouvant être admise au fonds de sécurité au même rang que des titres de sociétés suisses d'électricité, dont tous les avoirs se trouvent sur territoire suisse.

Regional newspaper, name unknown, Finhaut, Electricité d'Emosson Archives, 1971

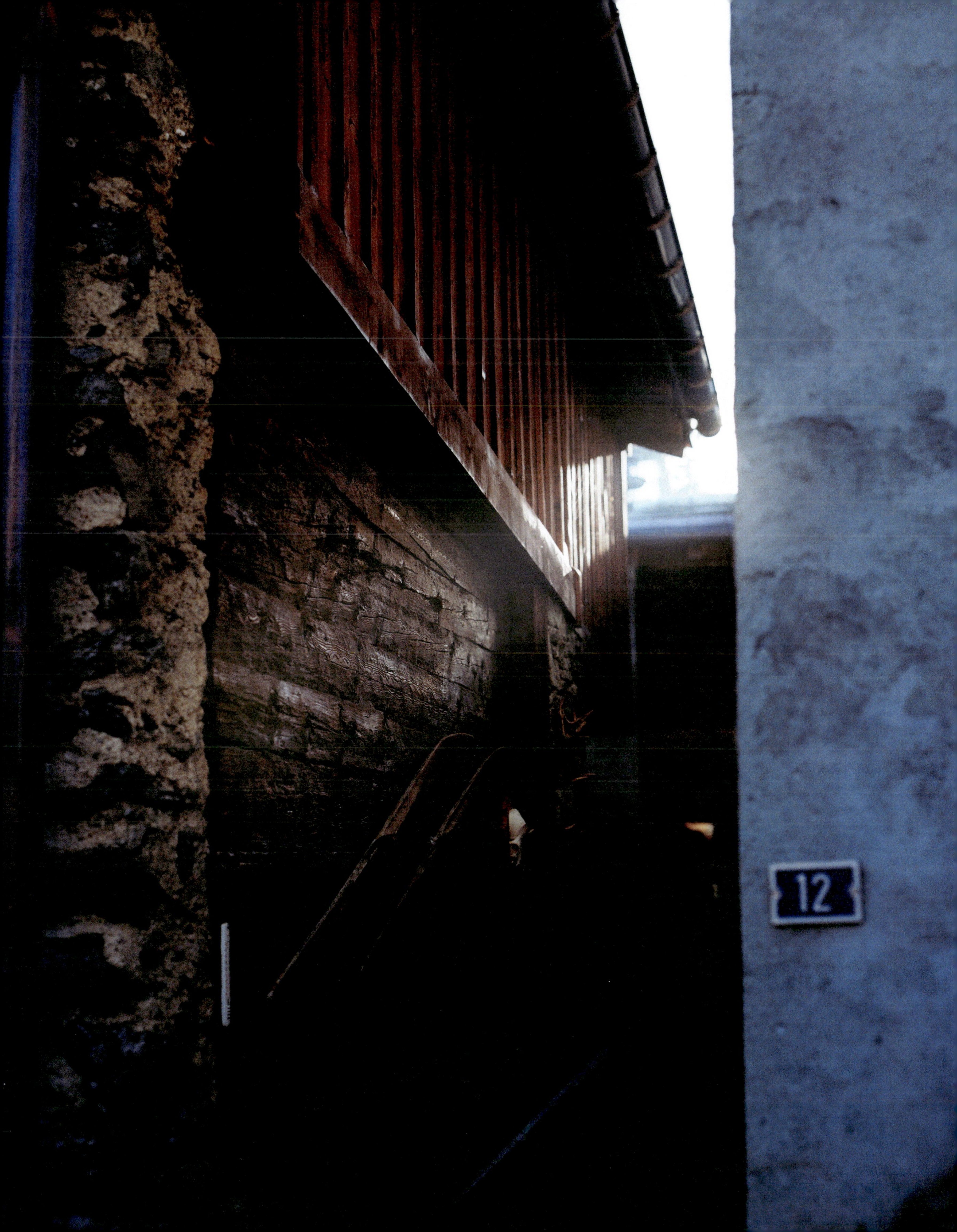
12

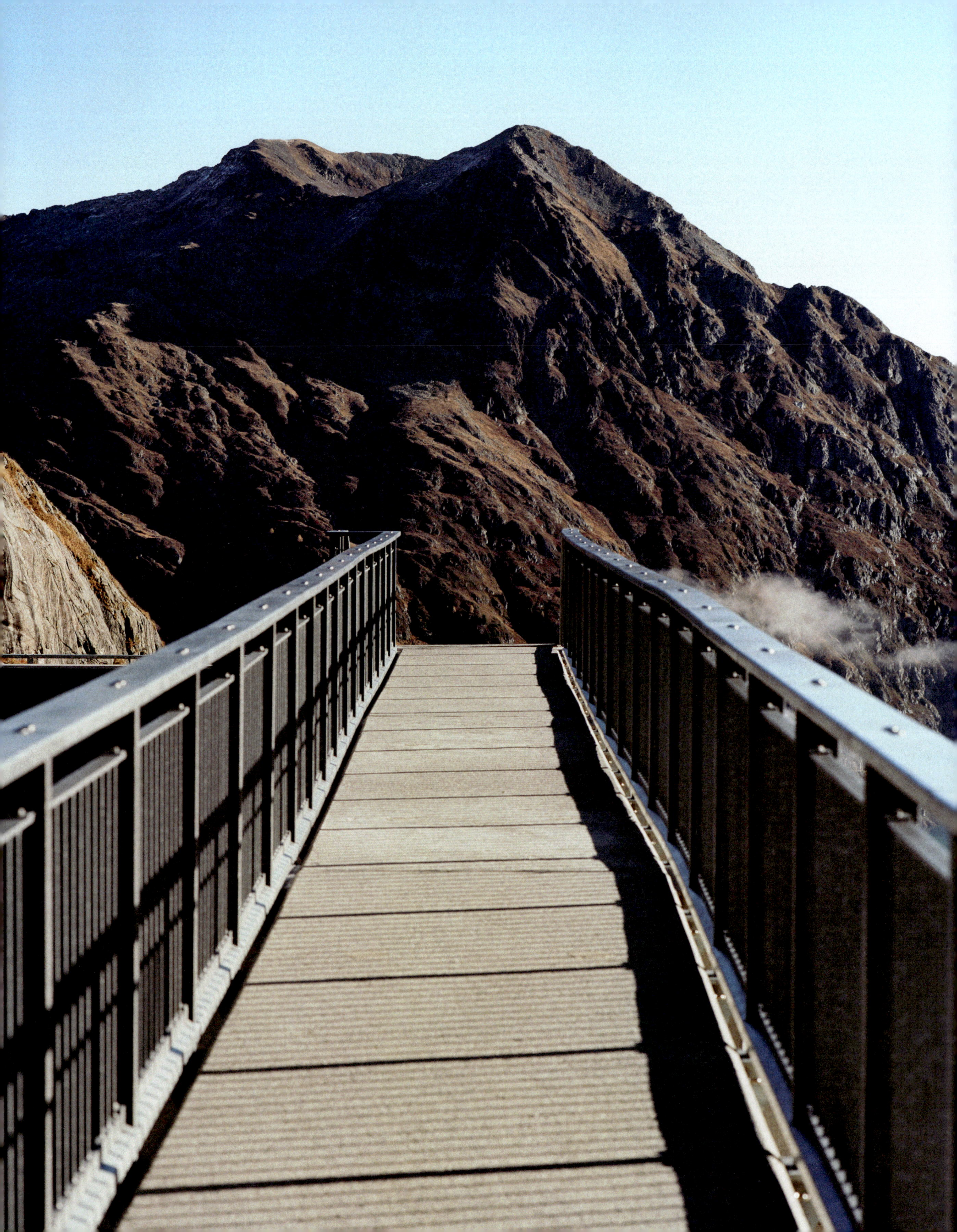

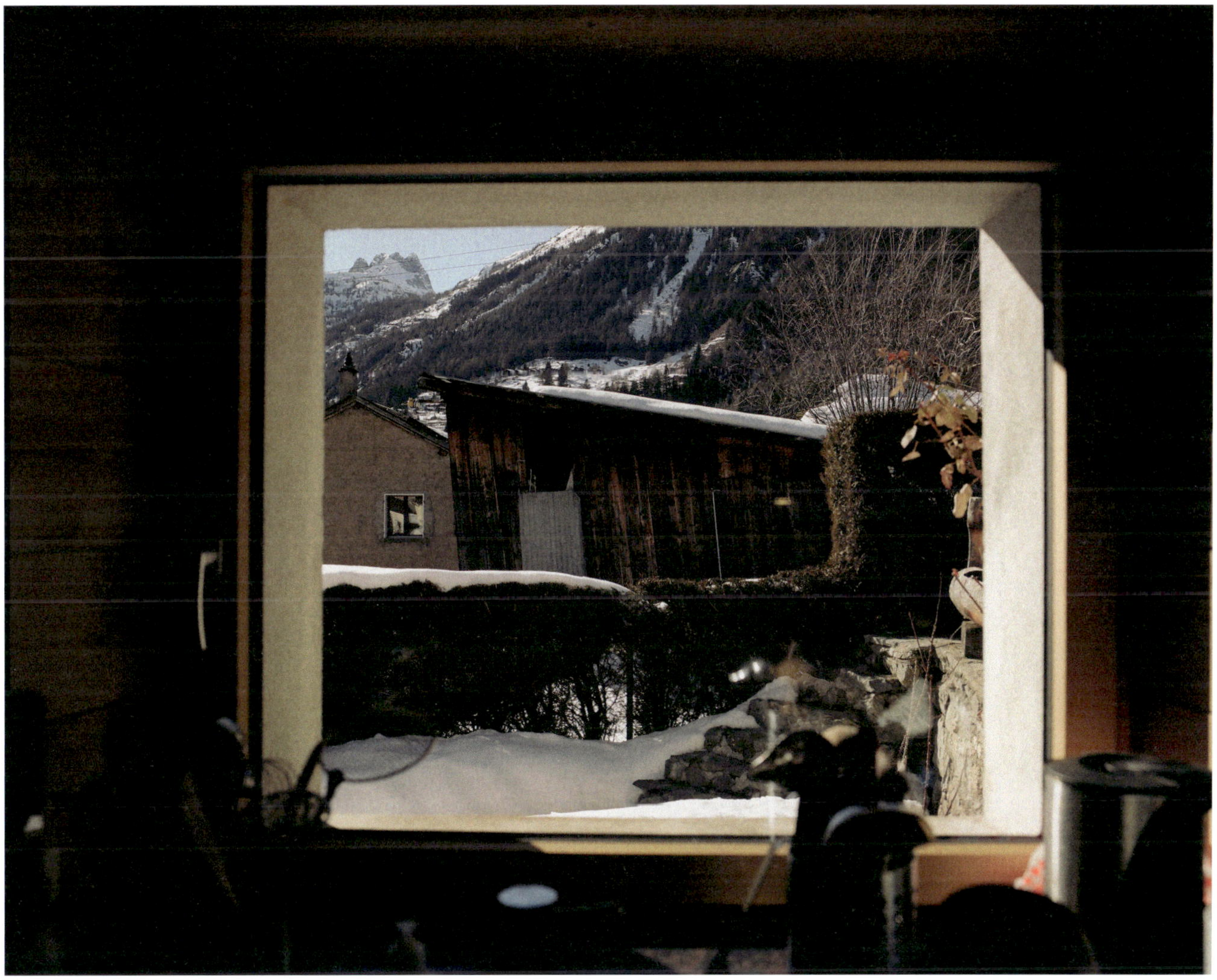

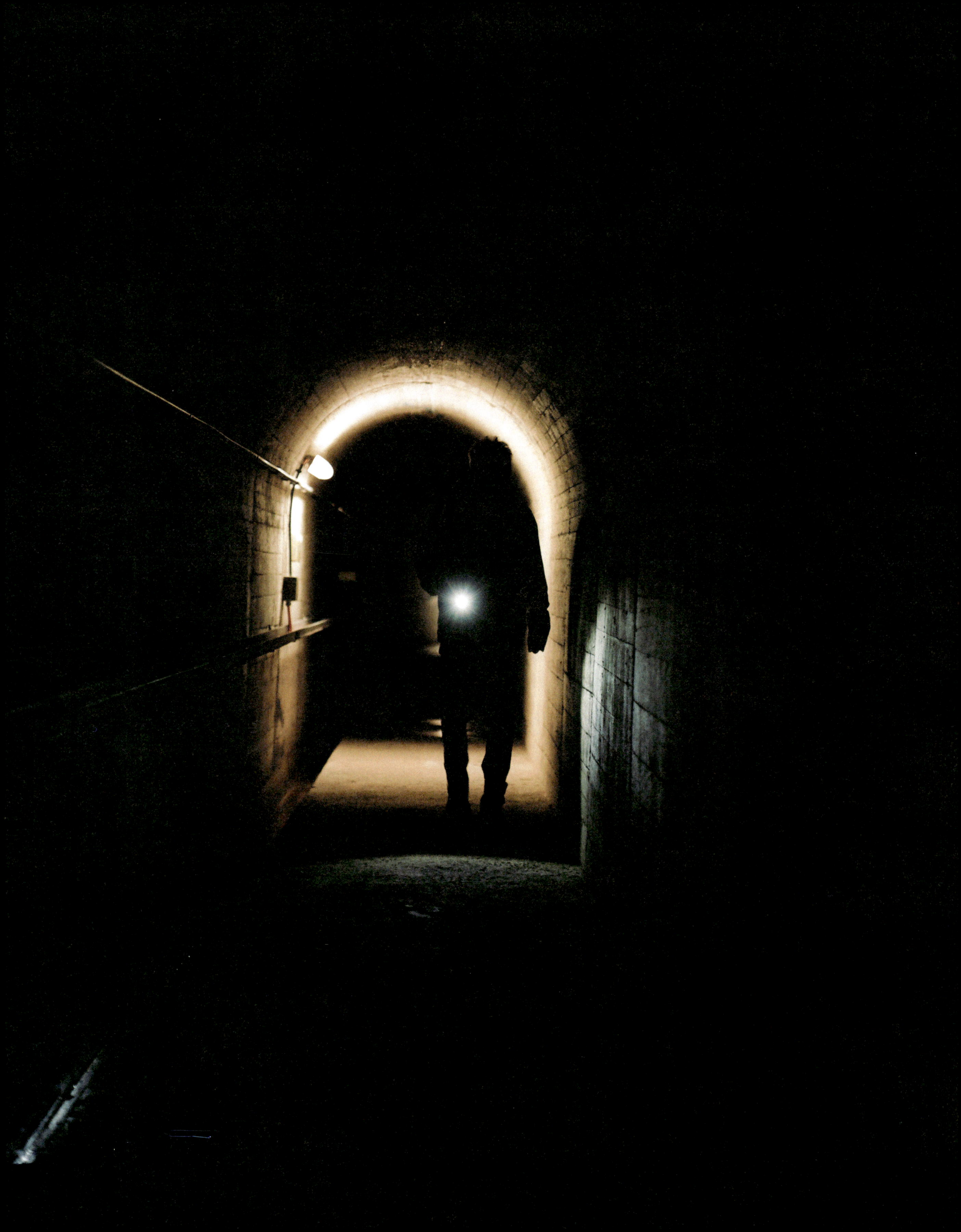

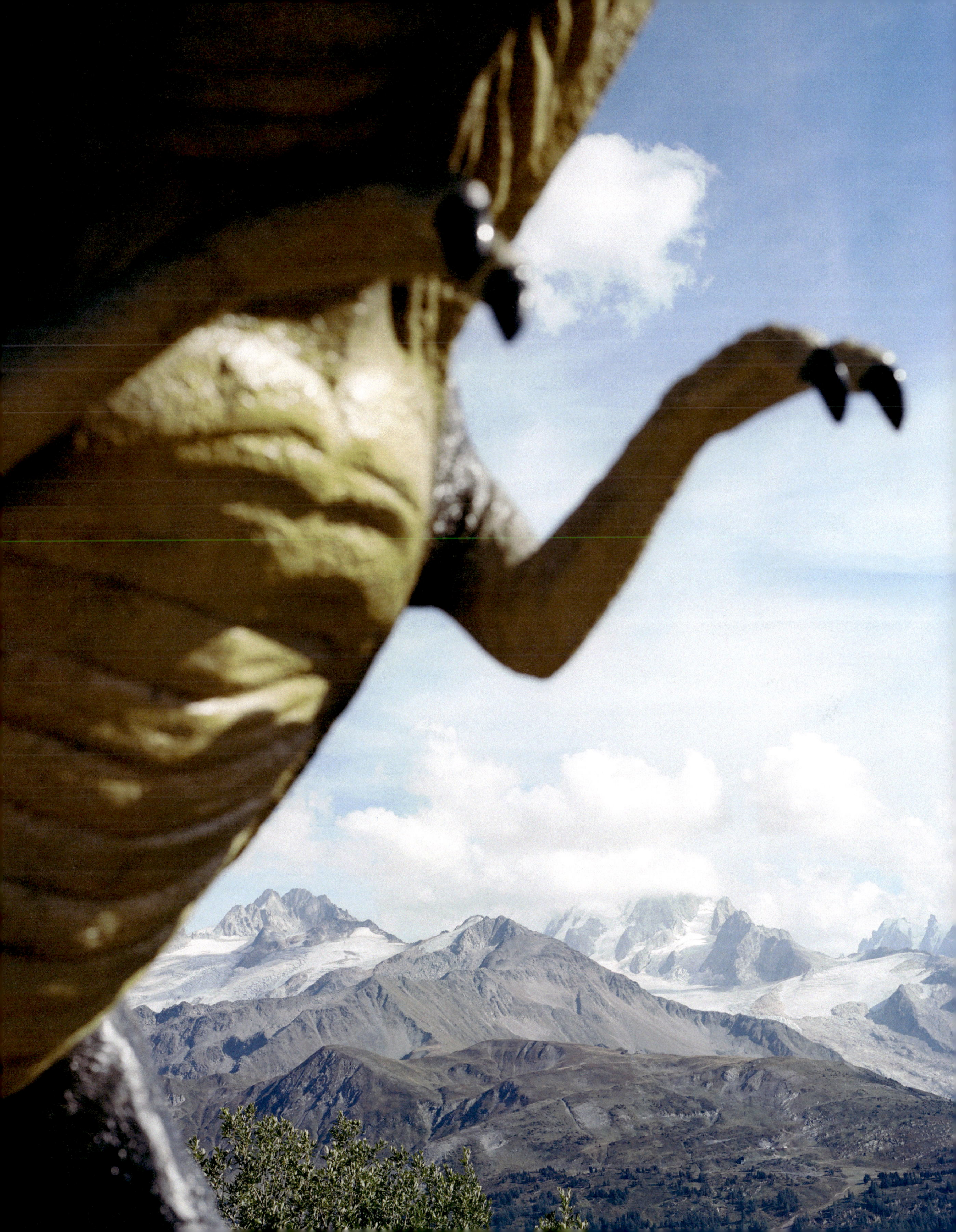

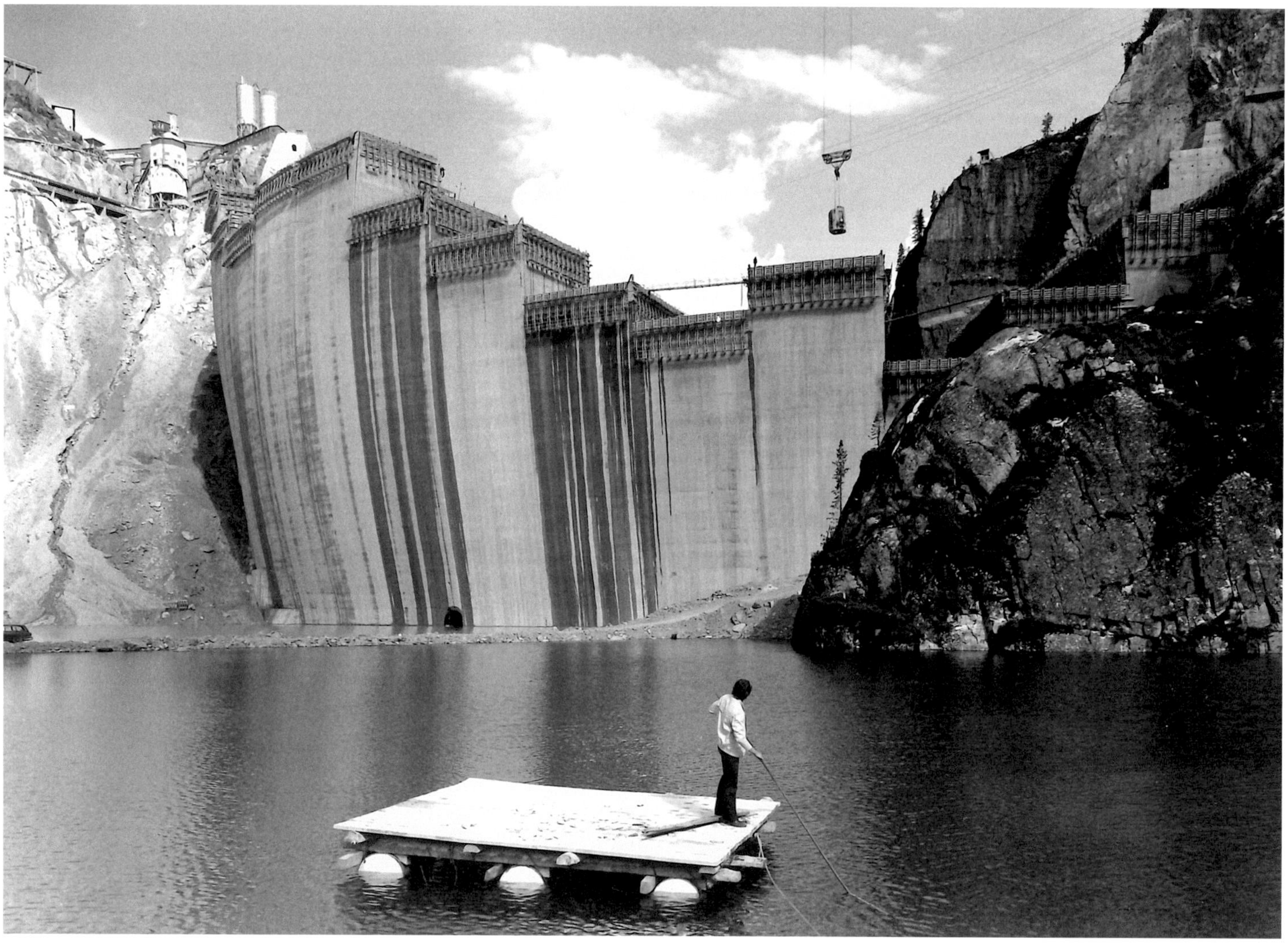

Dam, lake, and raft, Emosson dam,
Electricité d'Emosson SA,
M. Darbellay, 1972

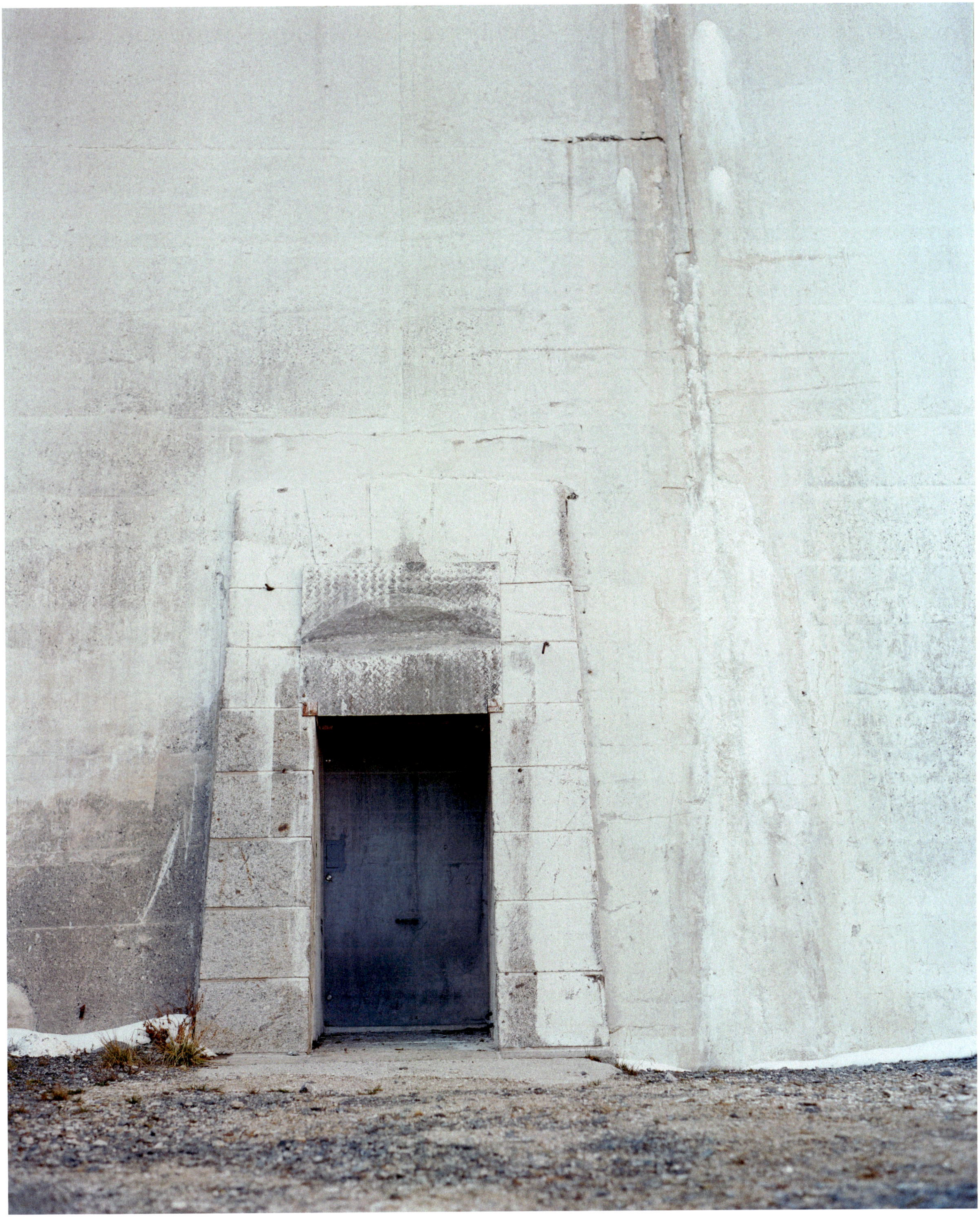

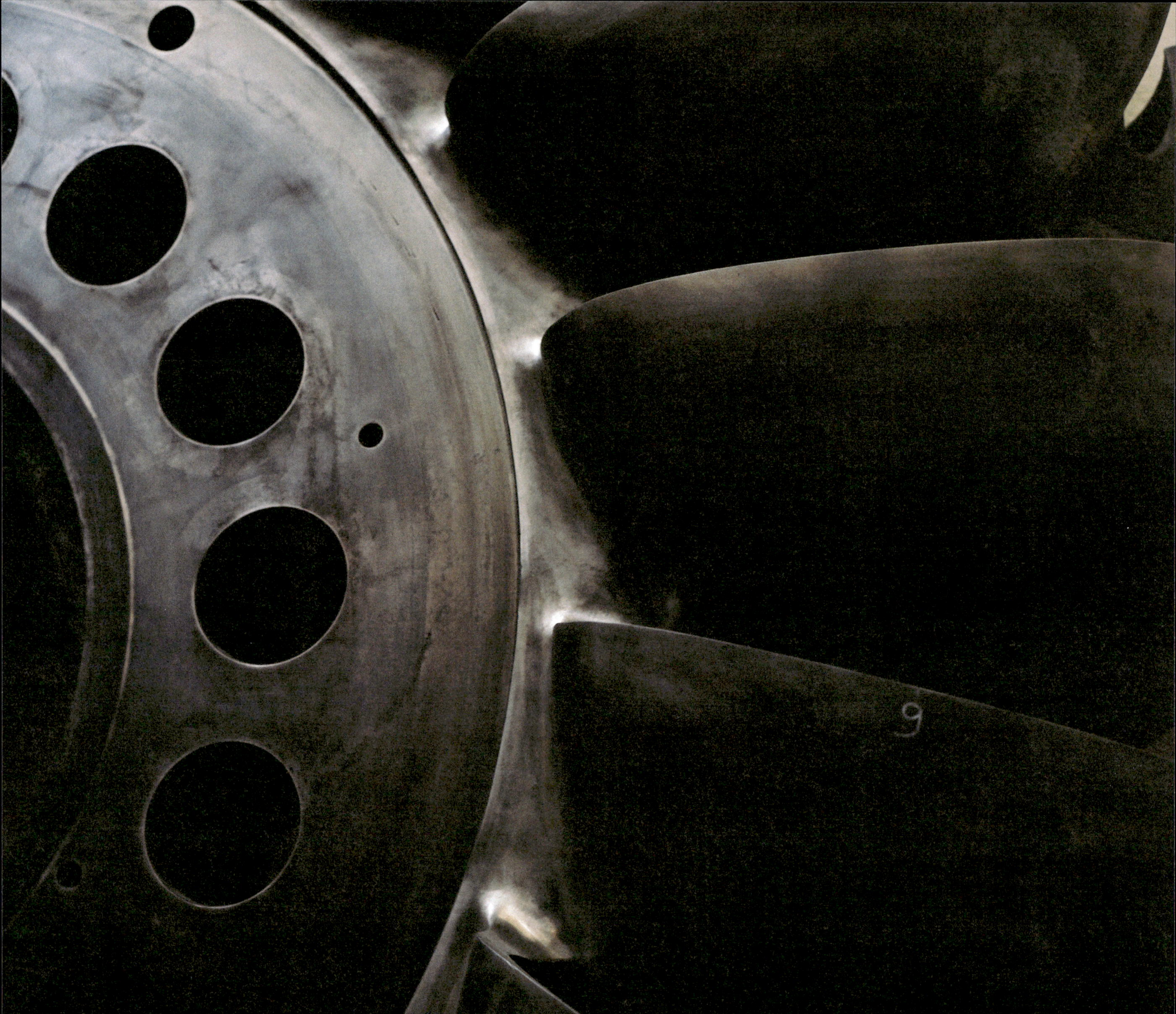

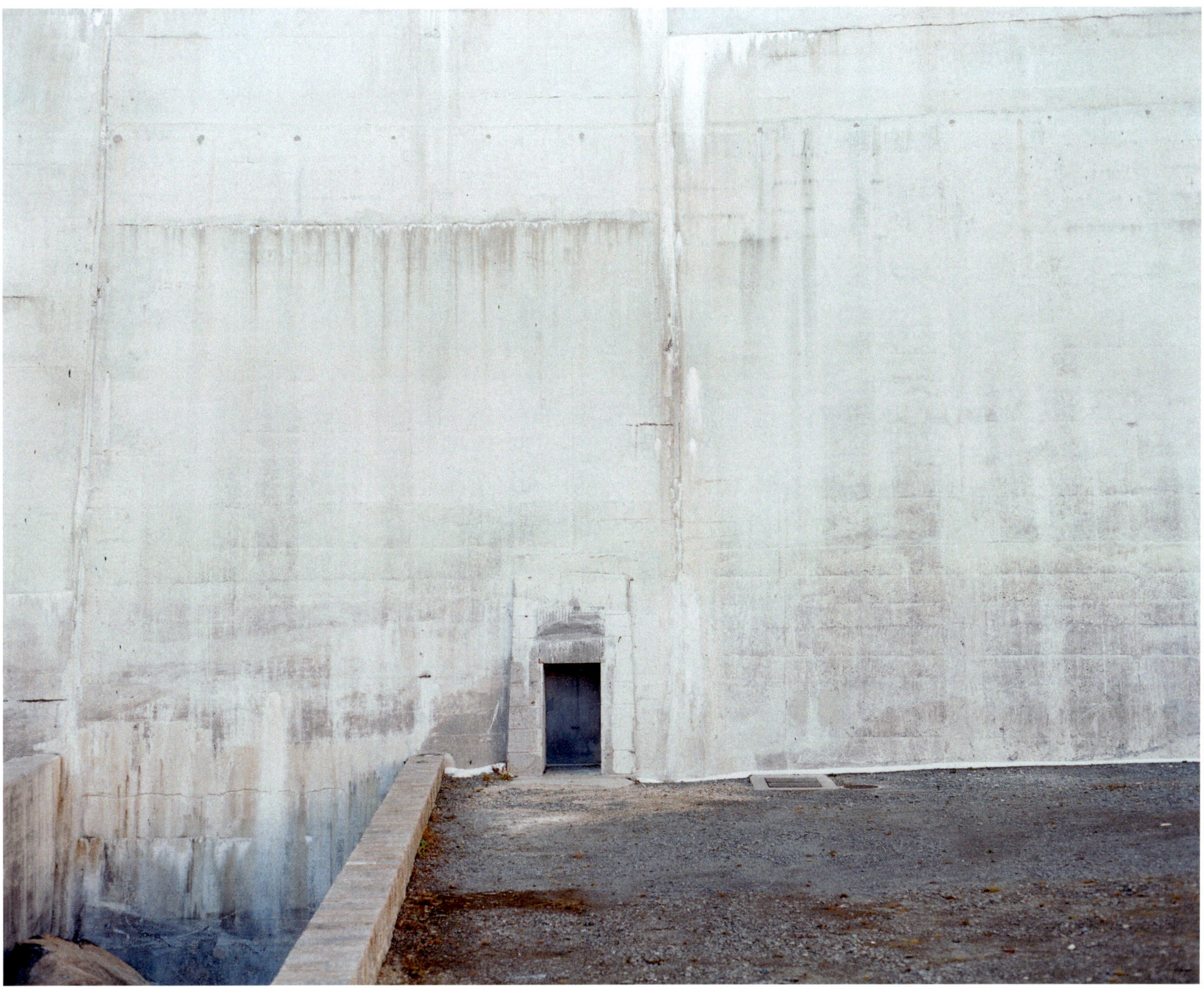

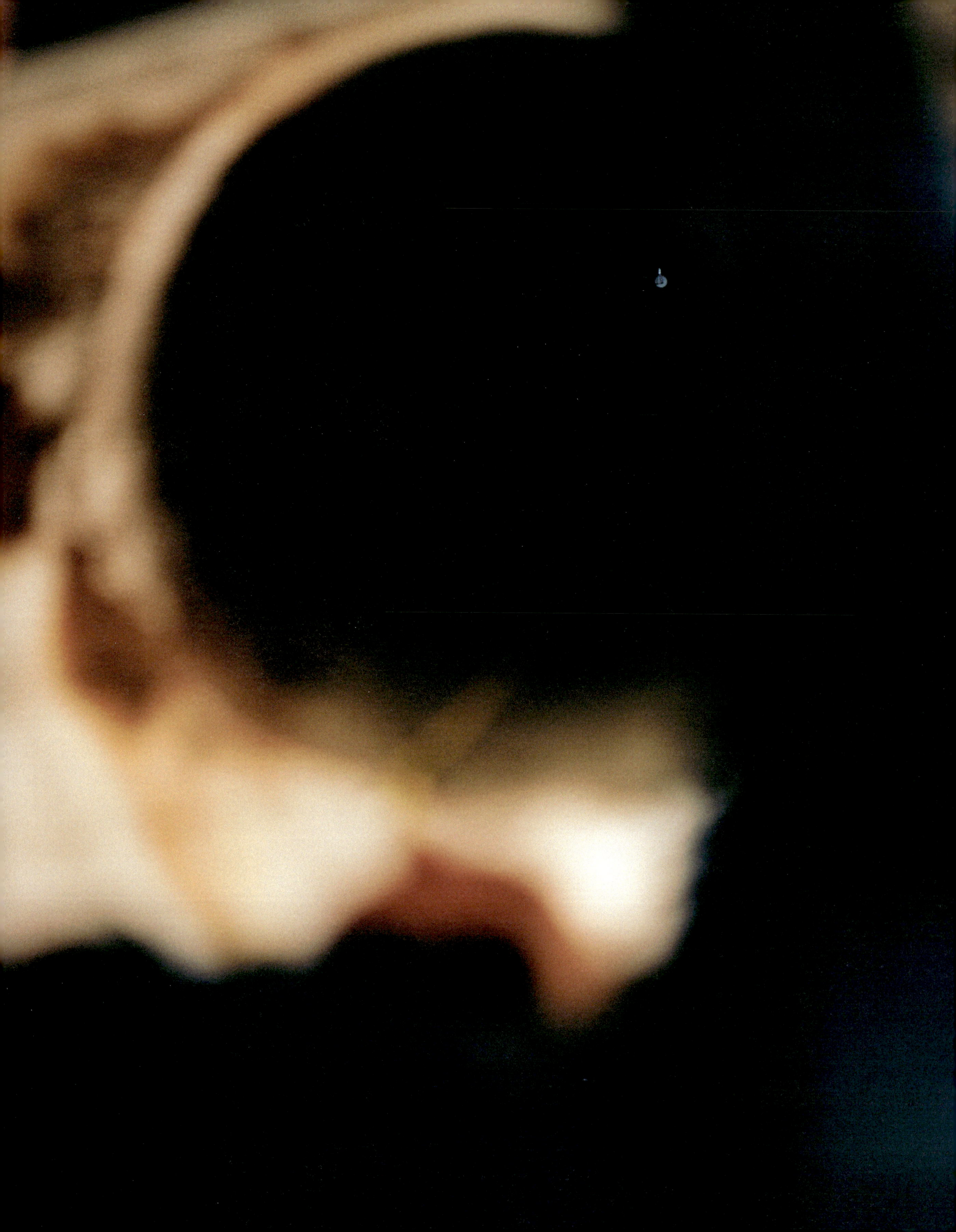

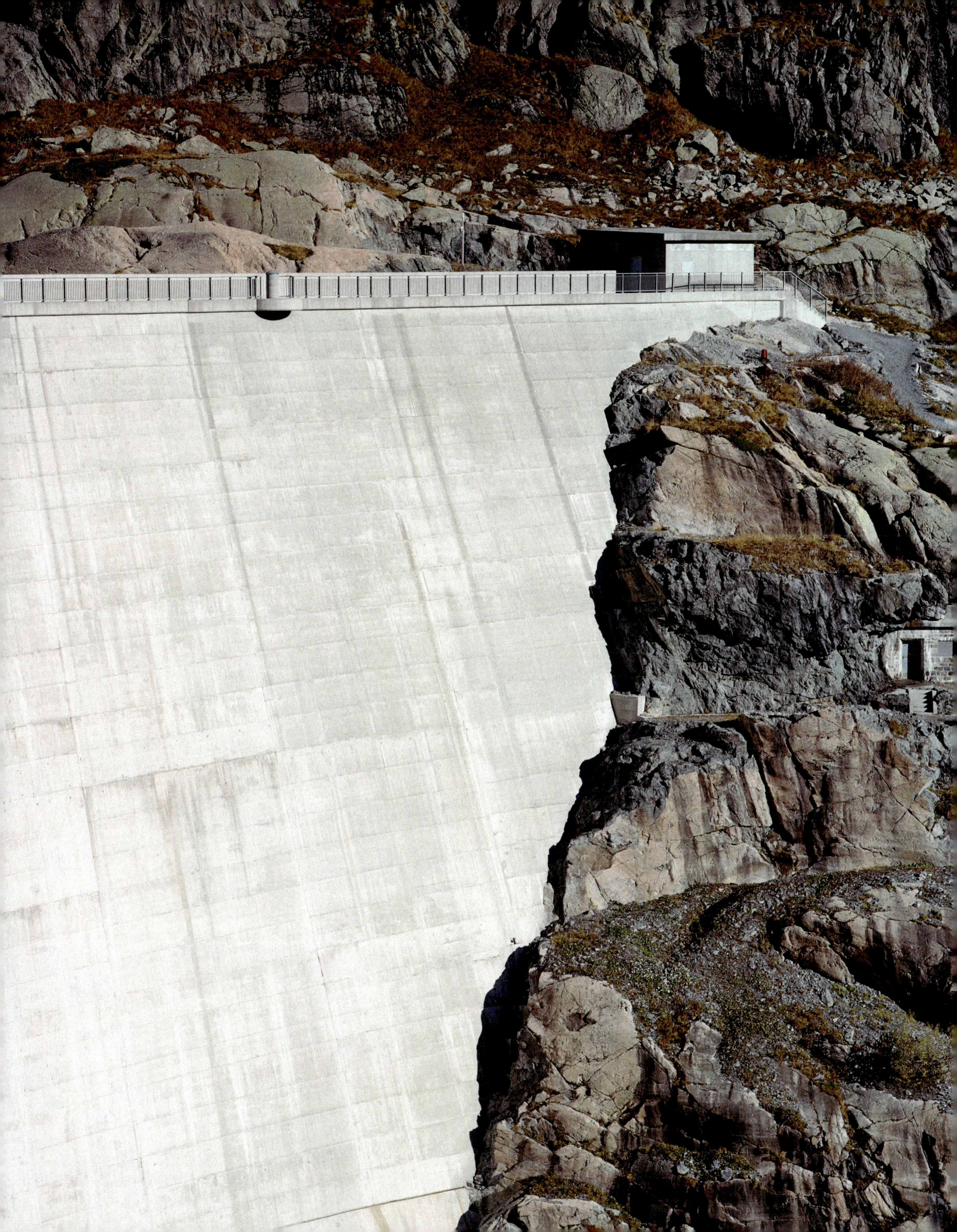

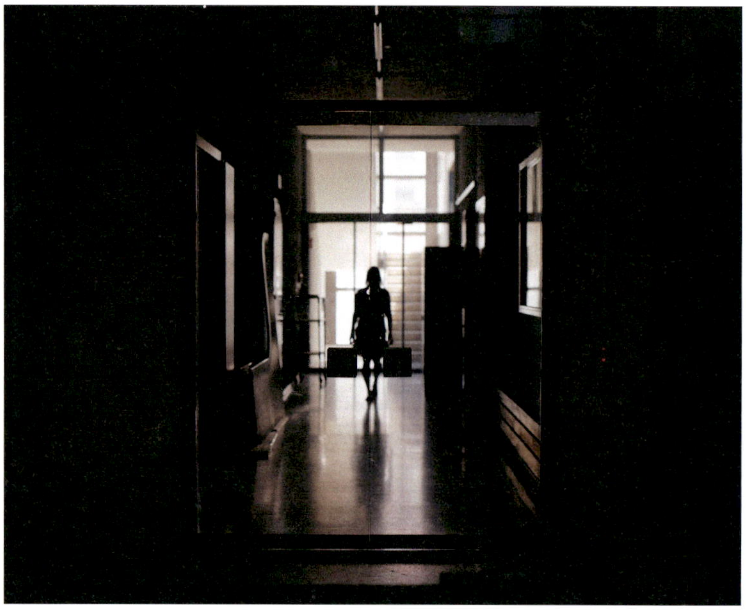

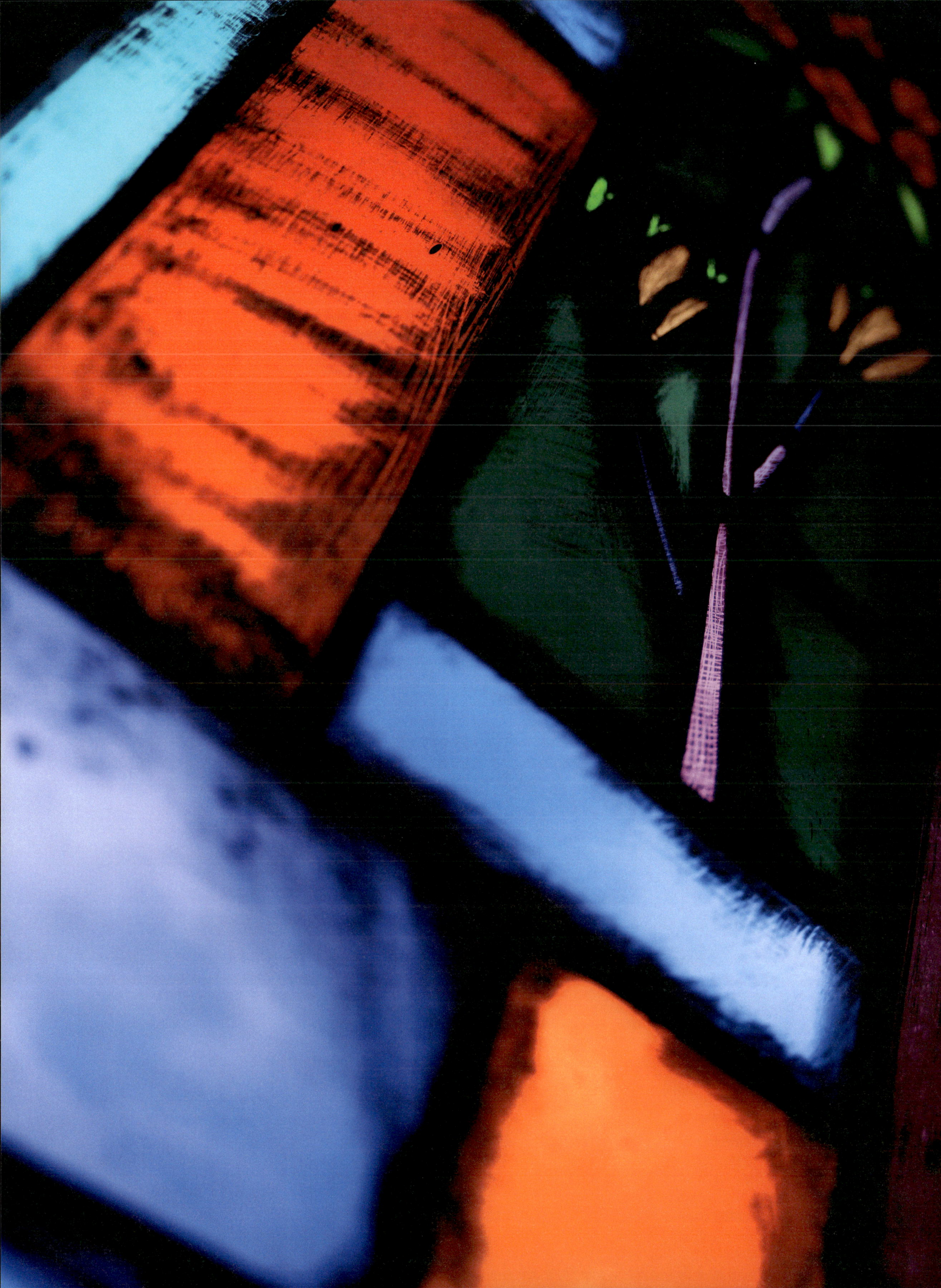

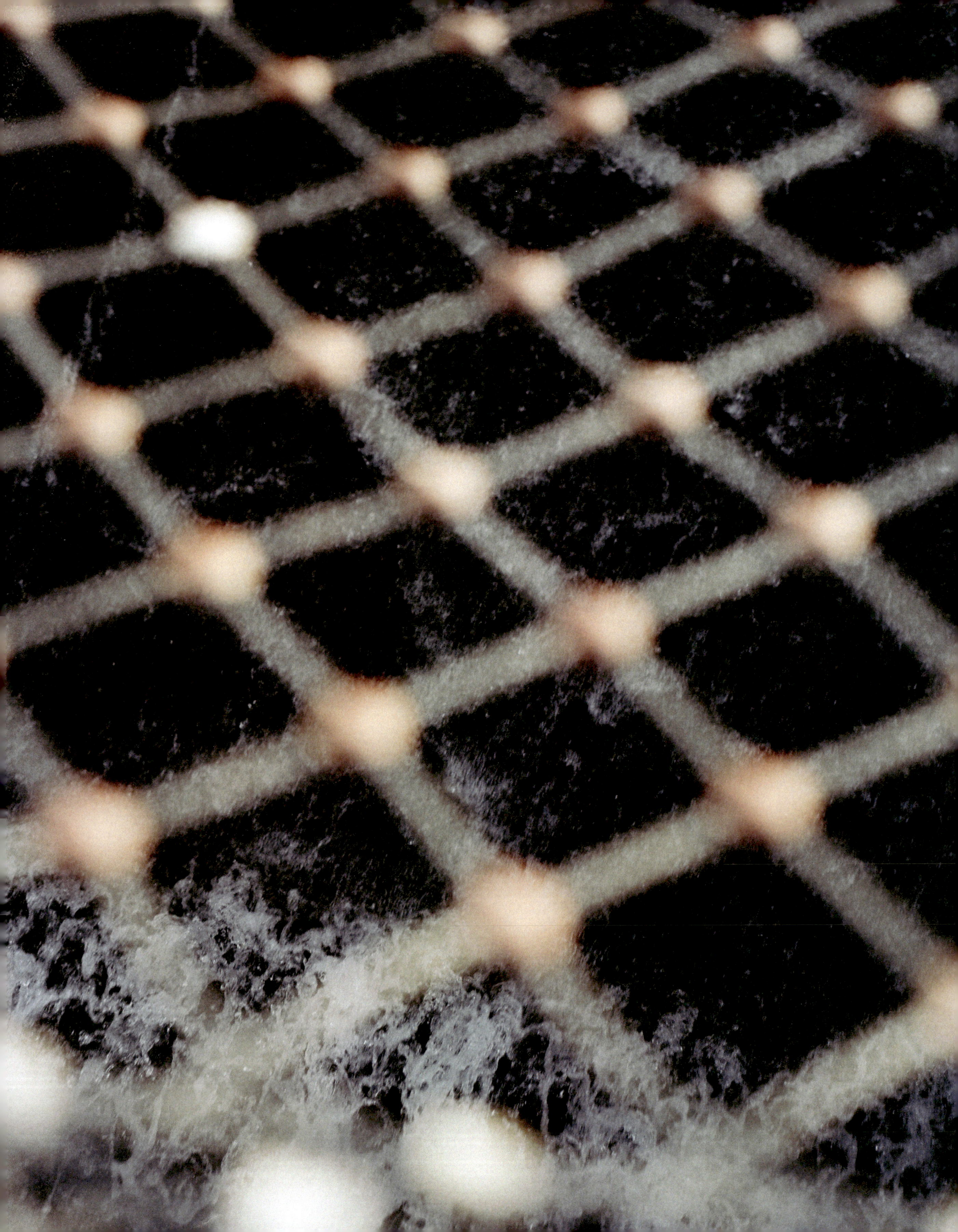

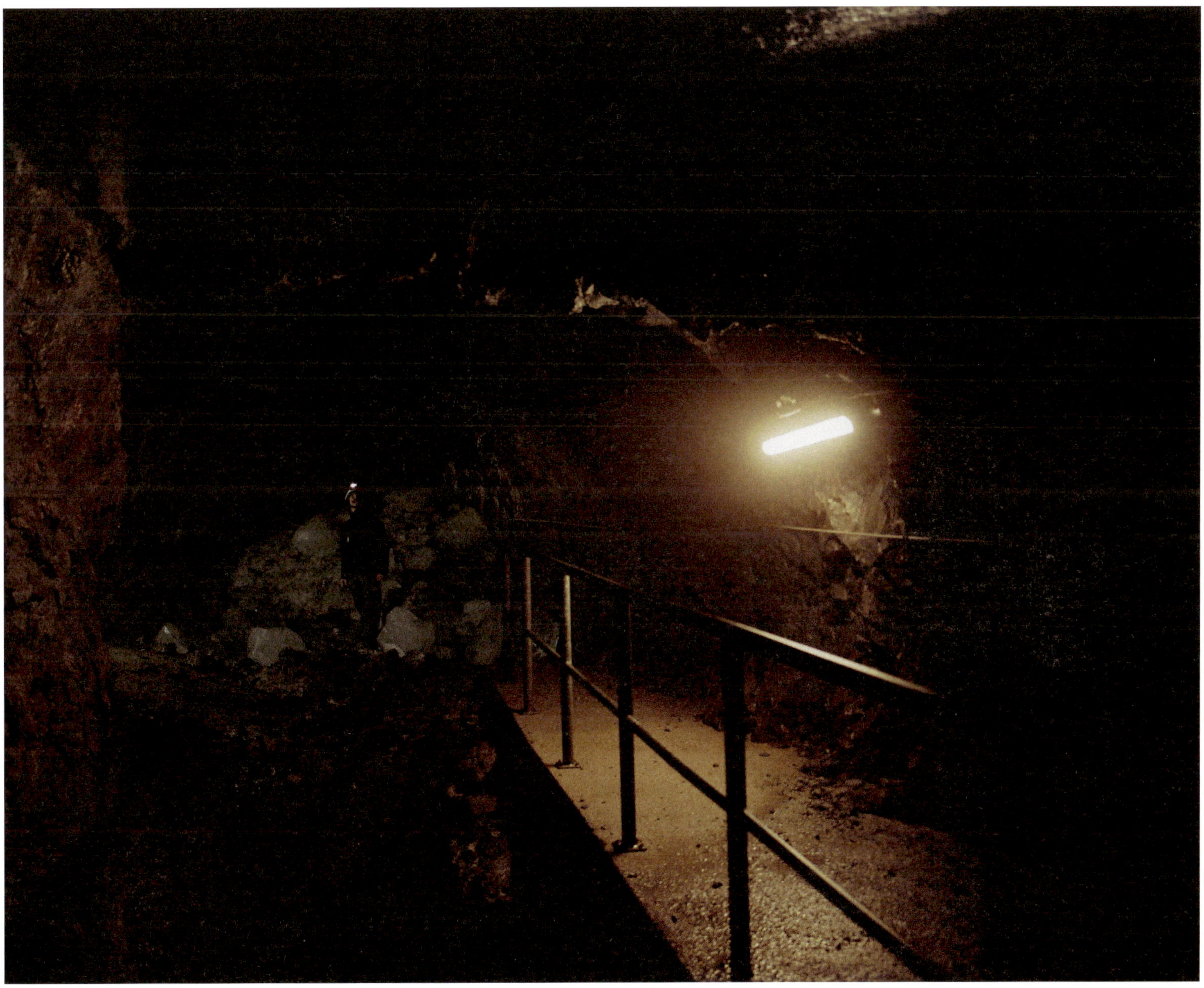

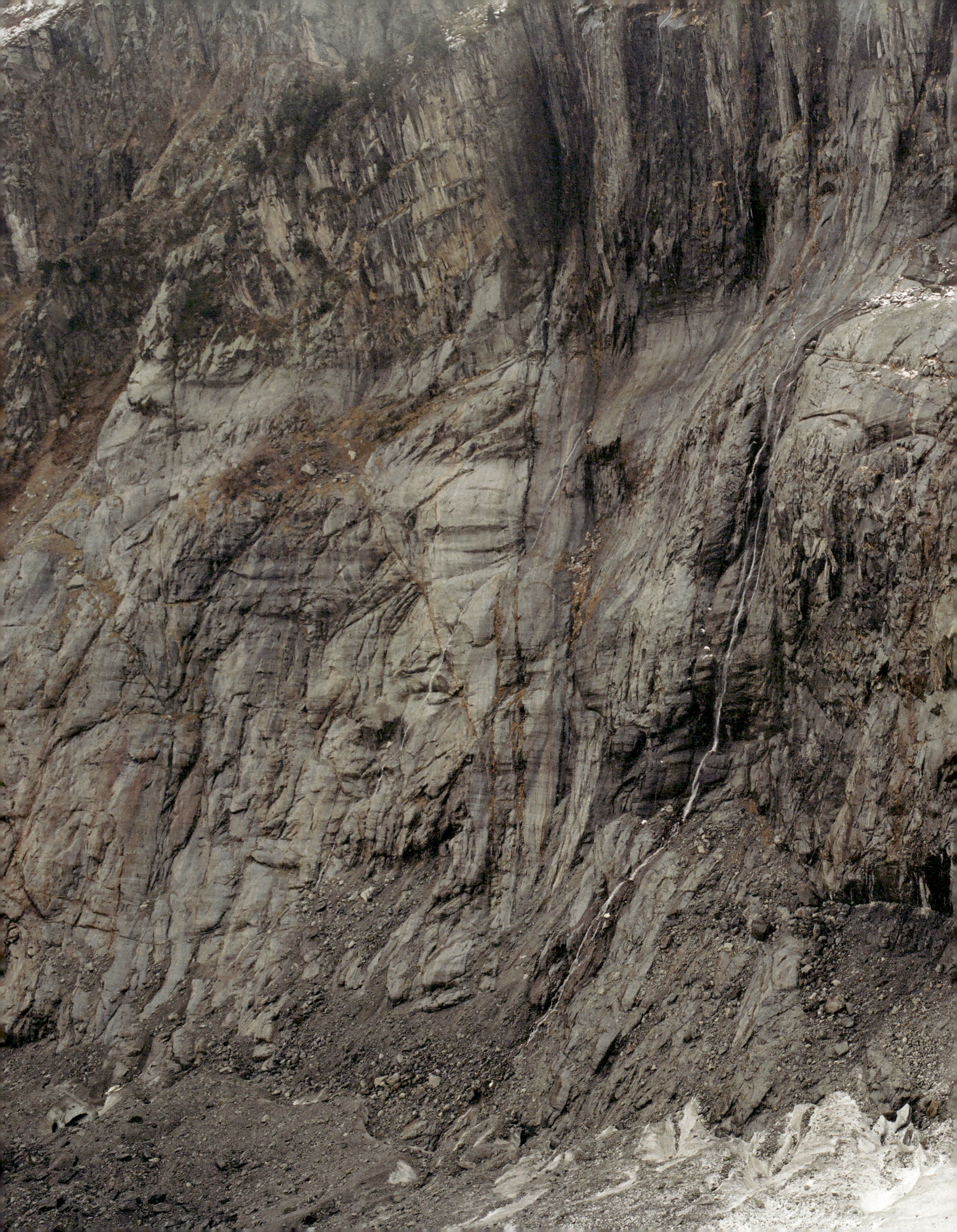

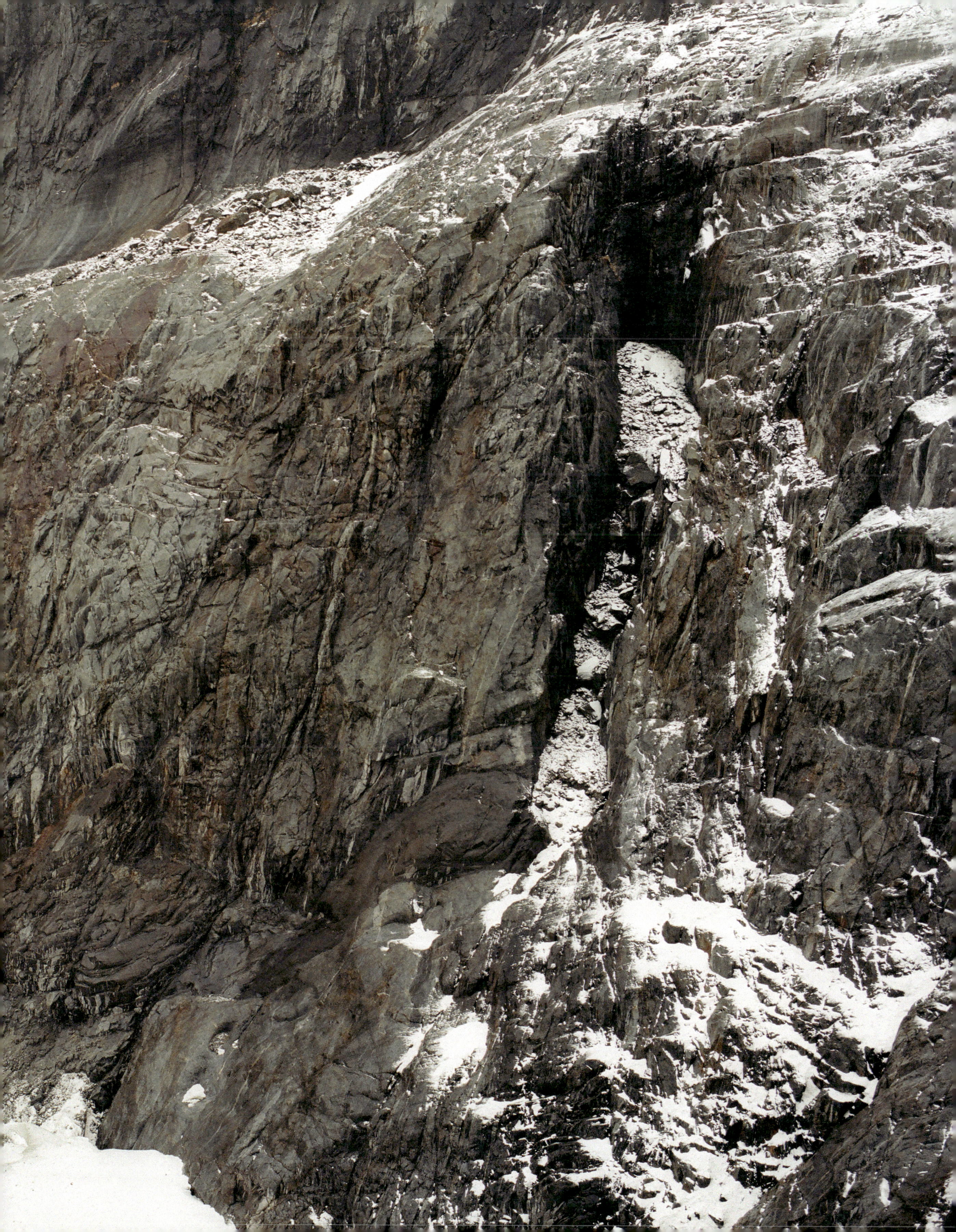

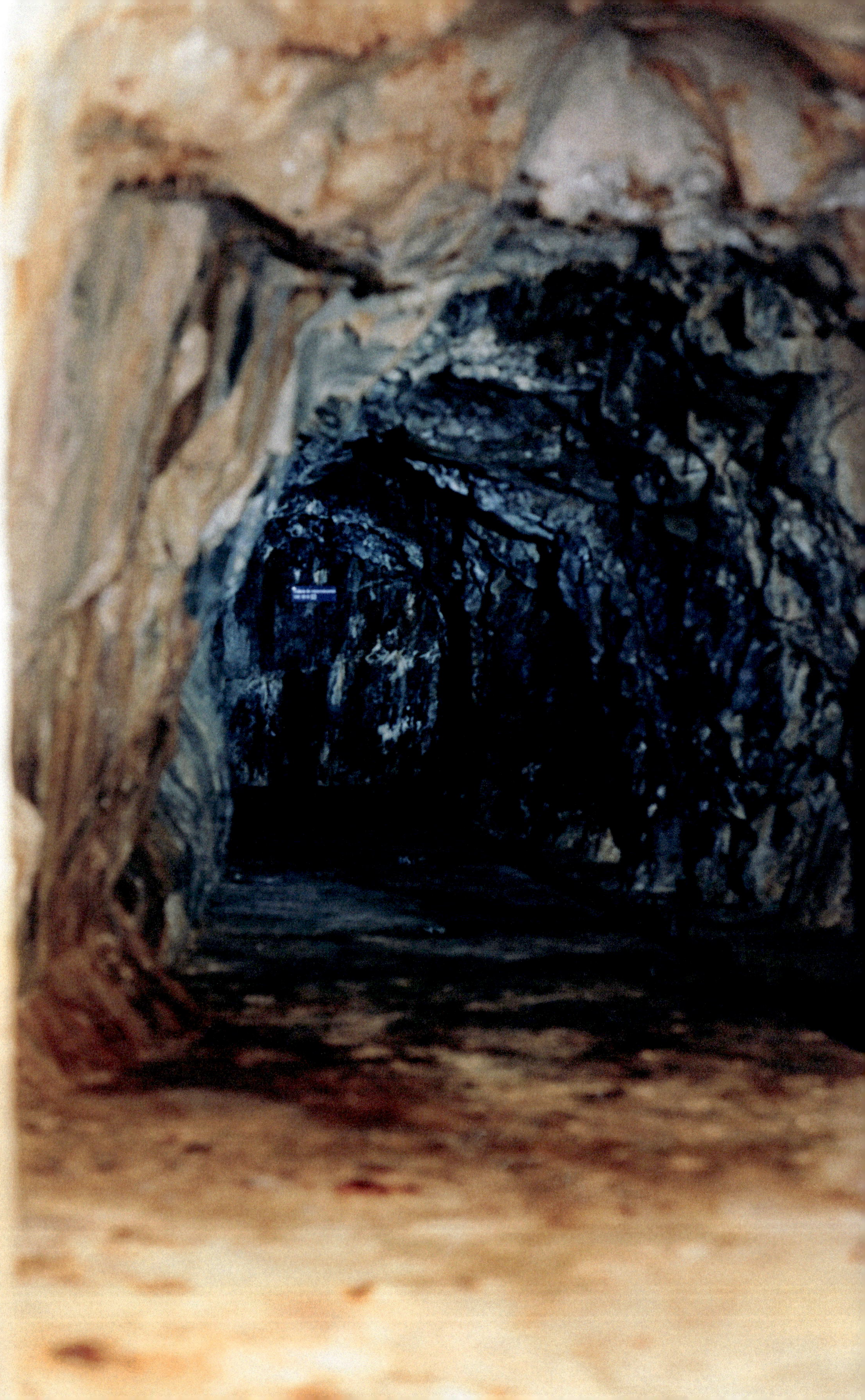

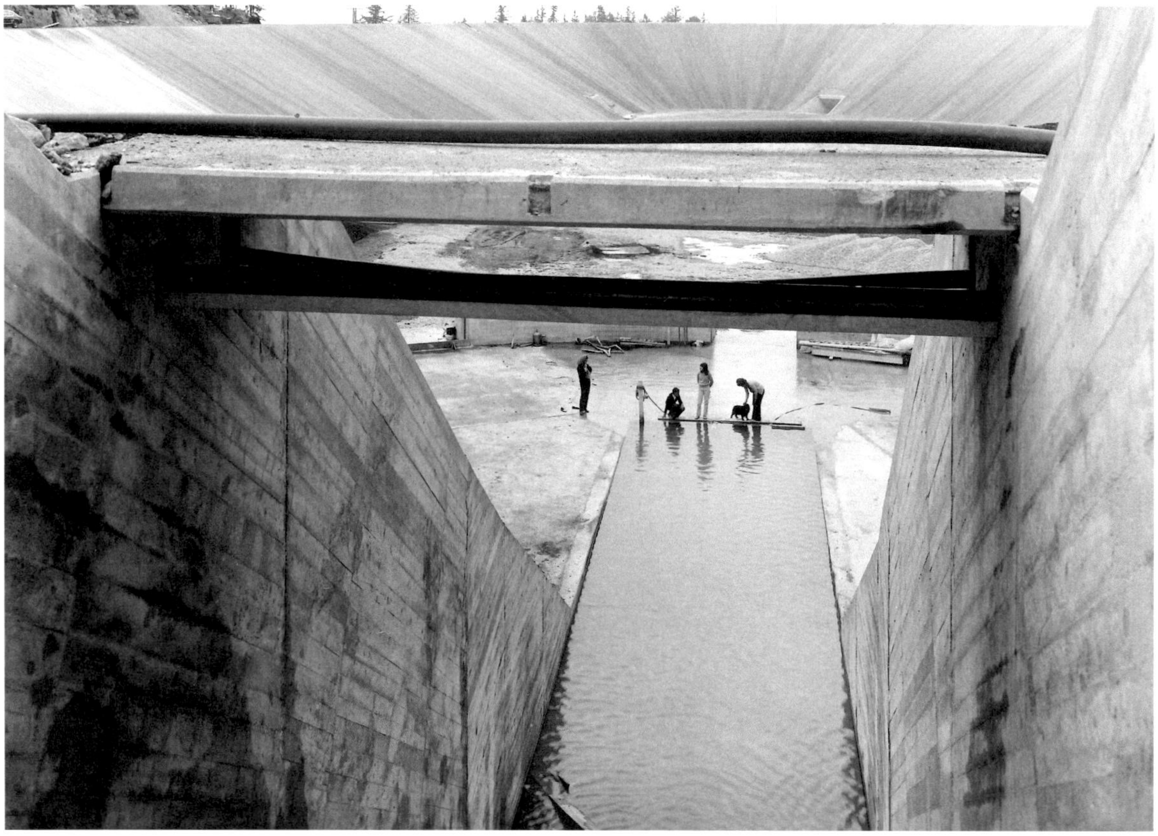

Basin des Esserts,
Electricité d'Emosson SA,
M. Darbellay, 1972

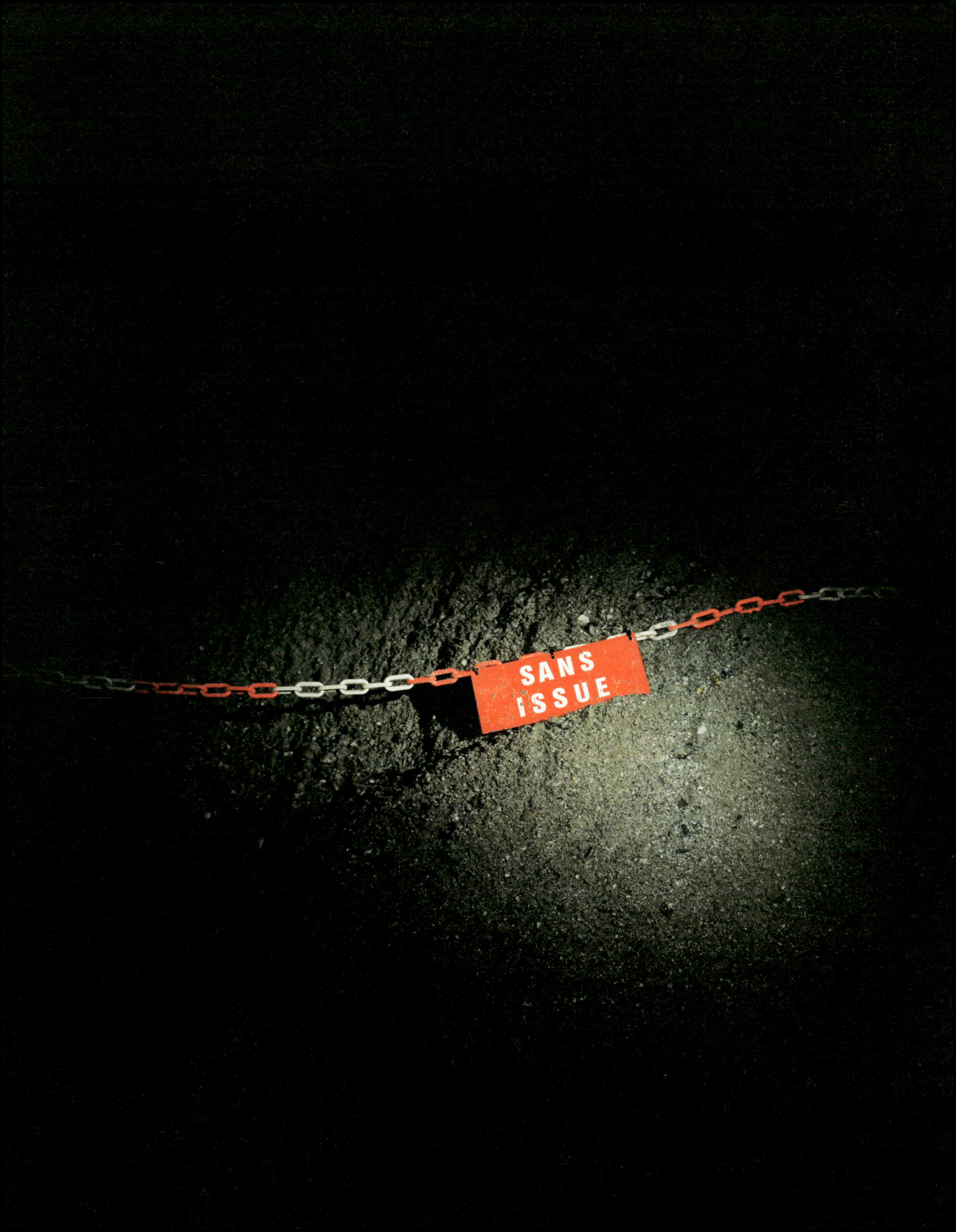
SANS
ISSUE

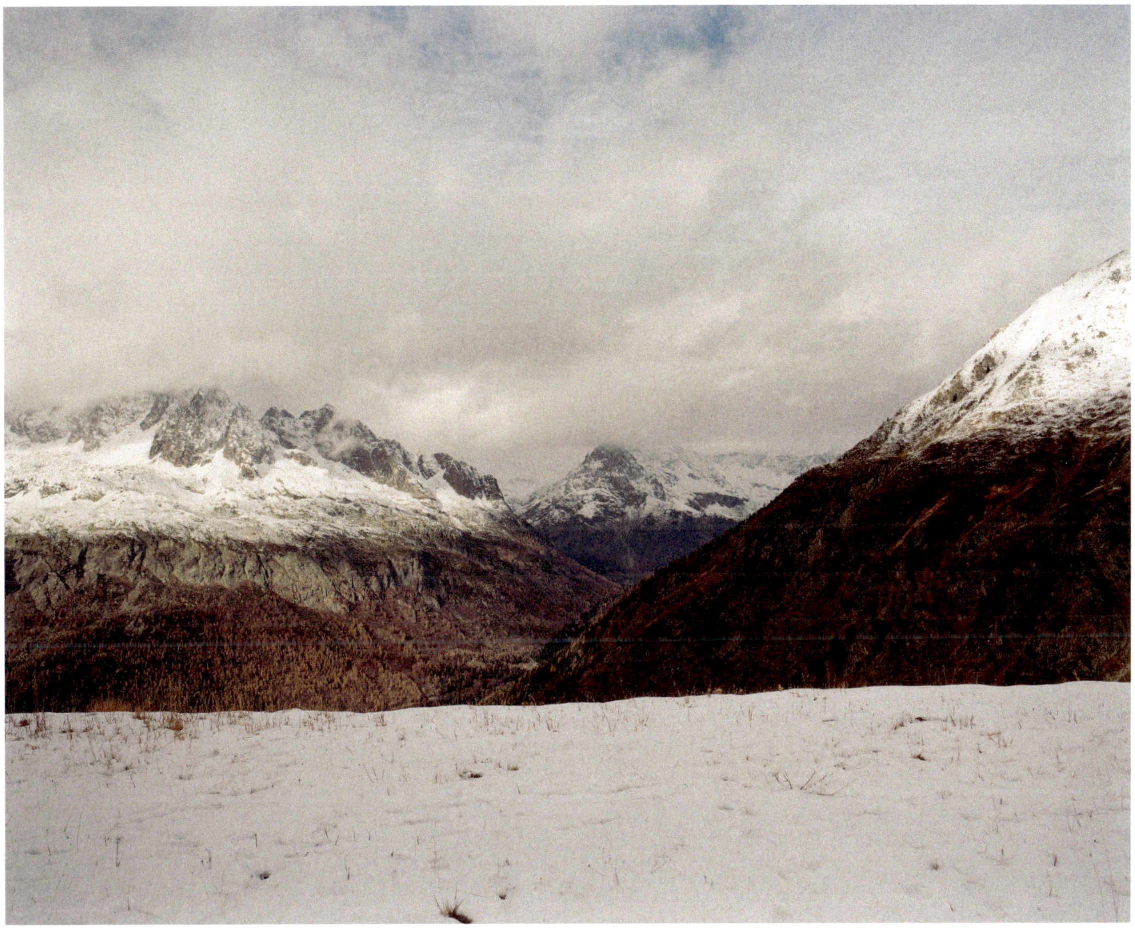

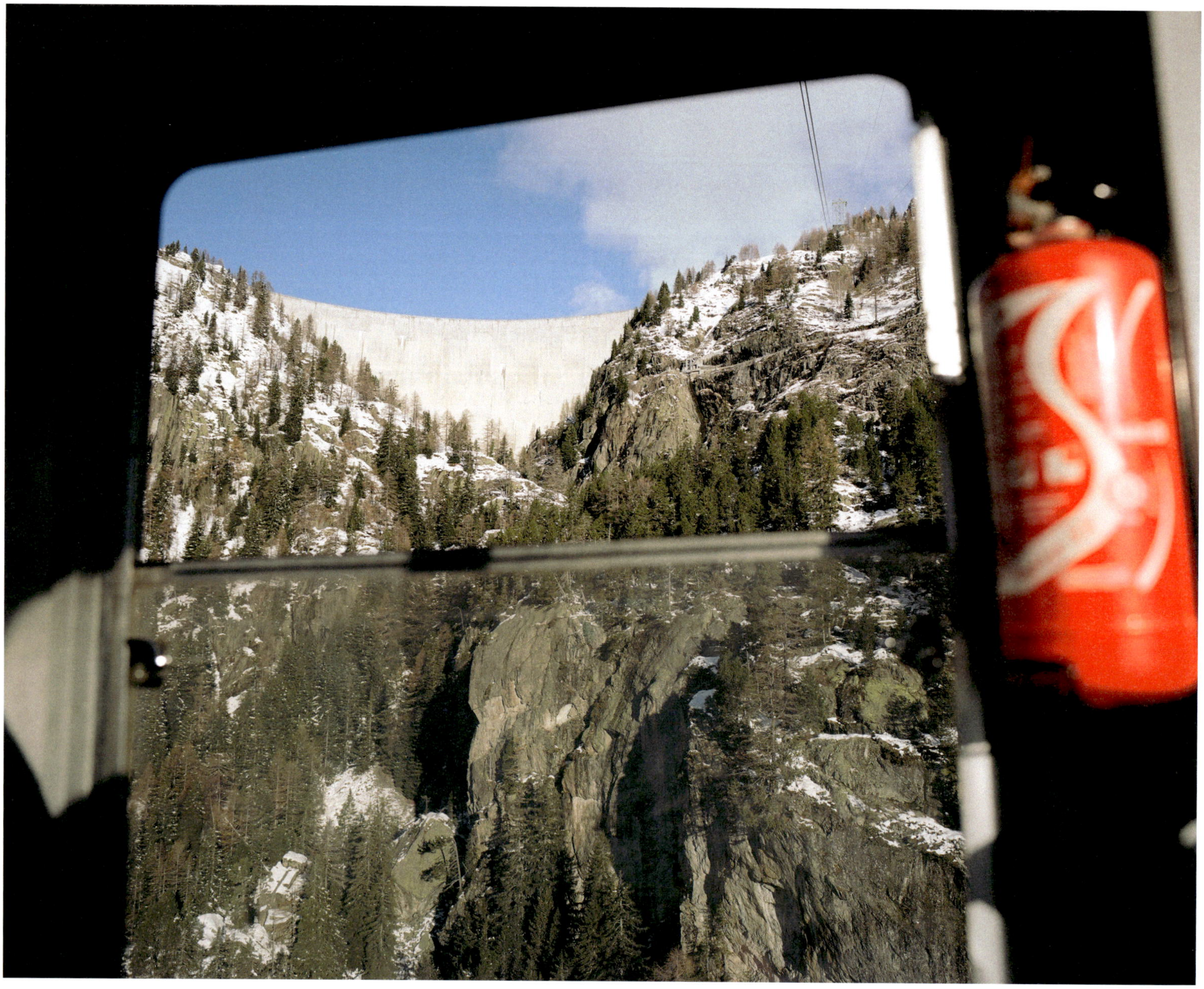

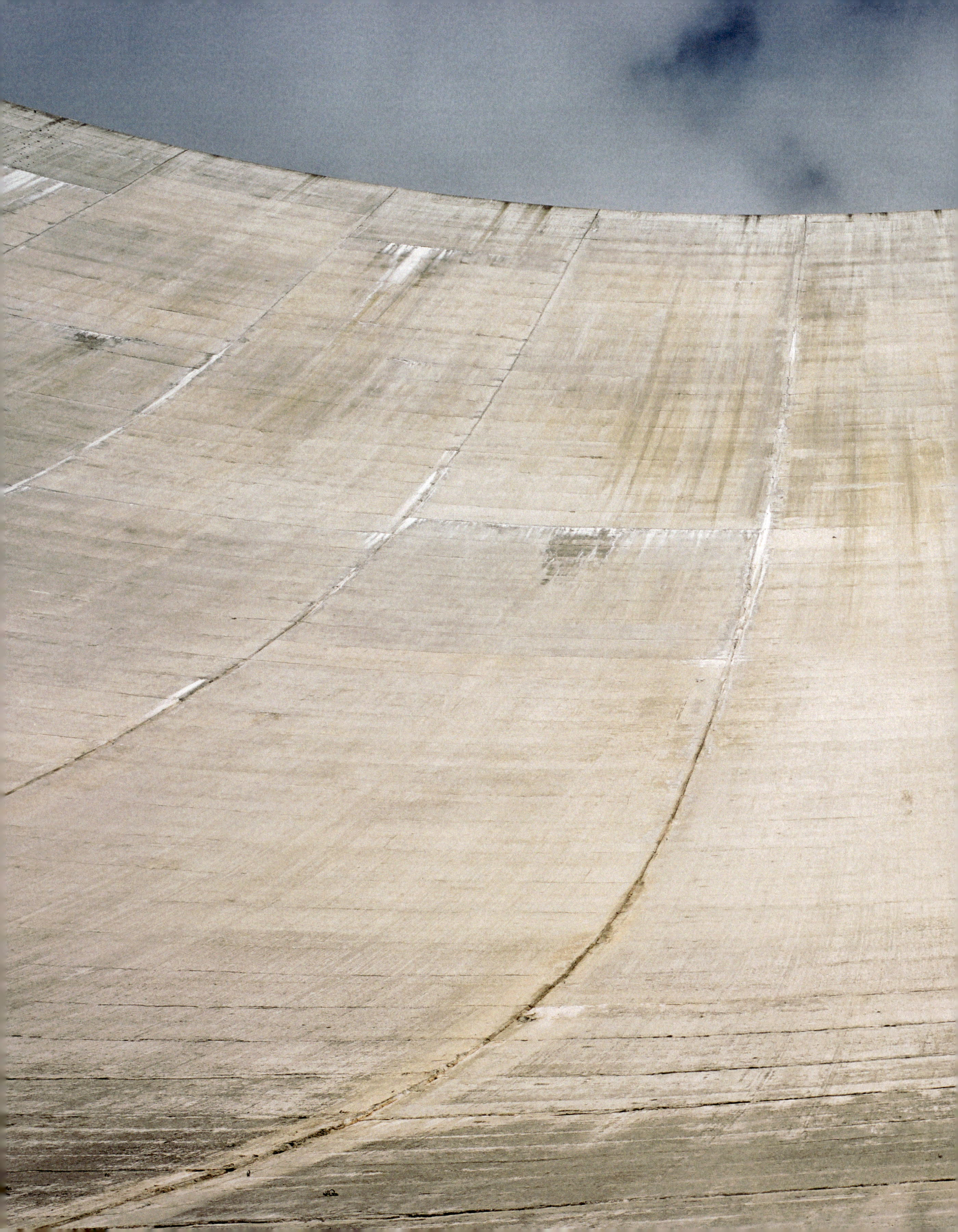

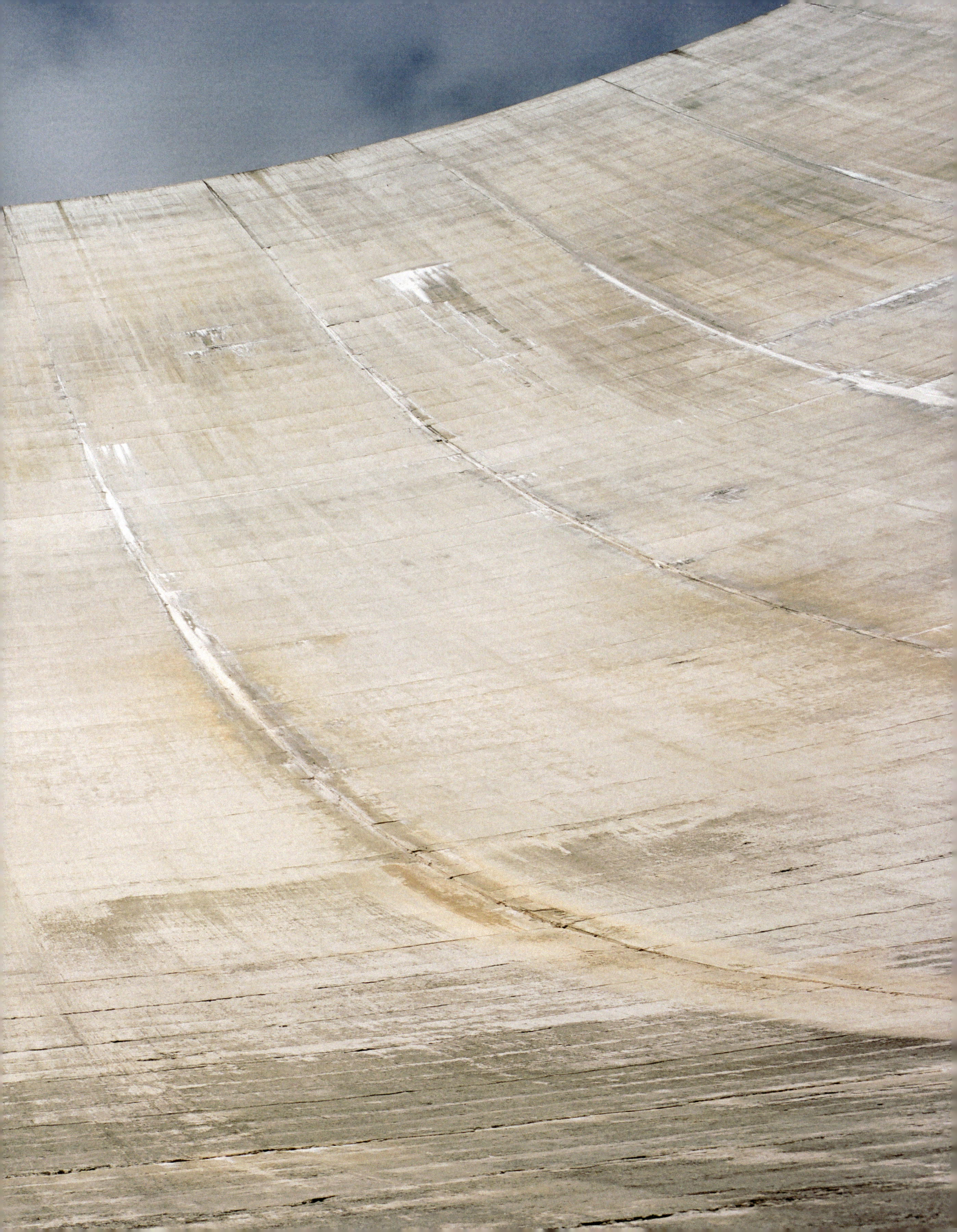

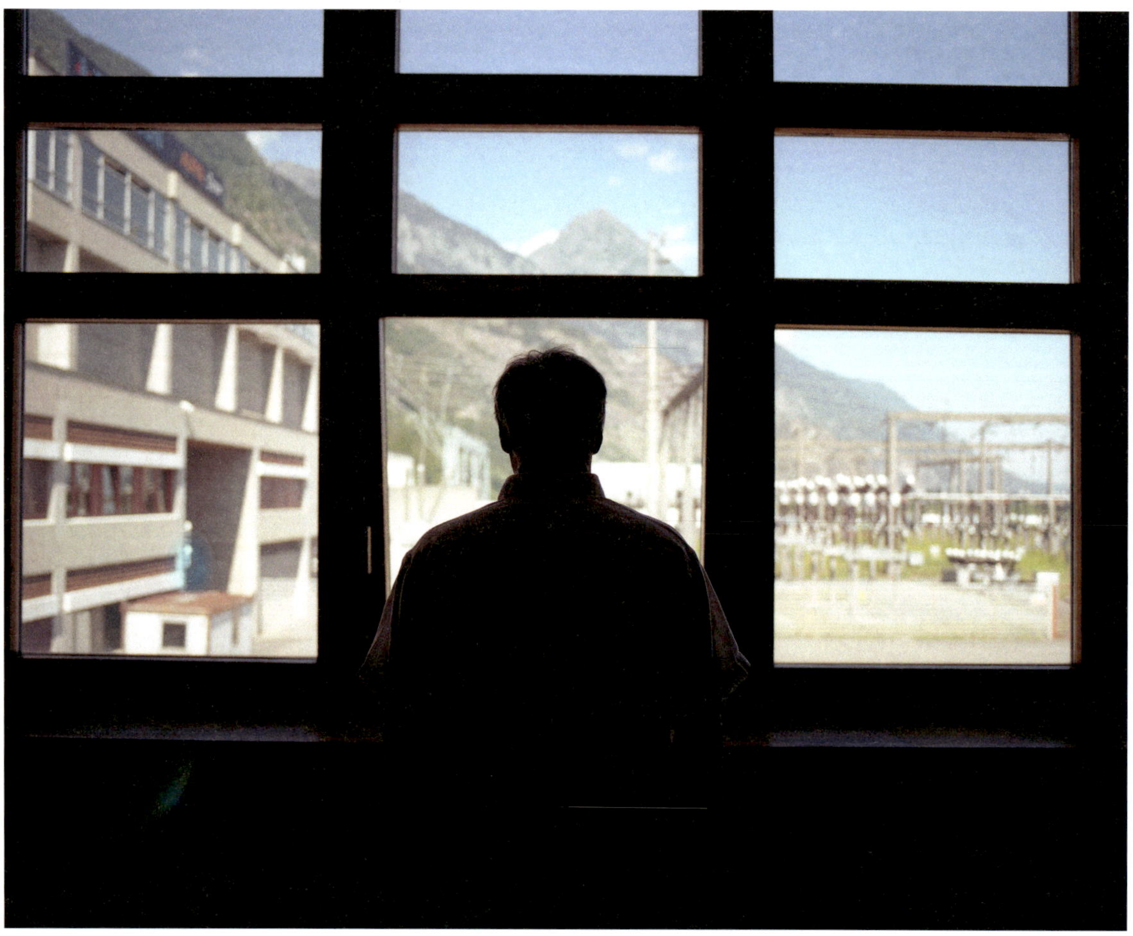

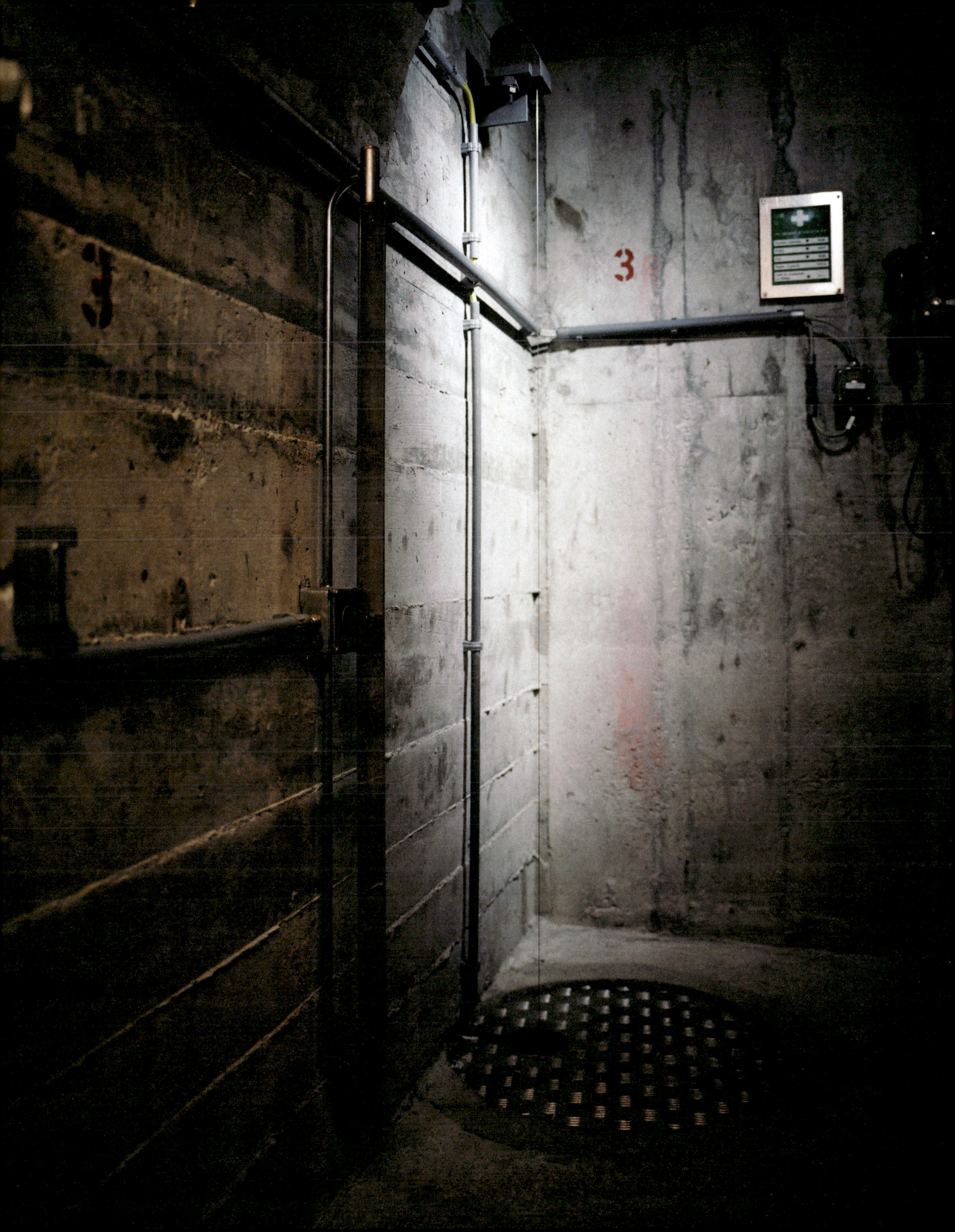
3

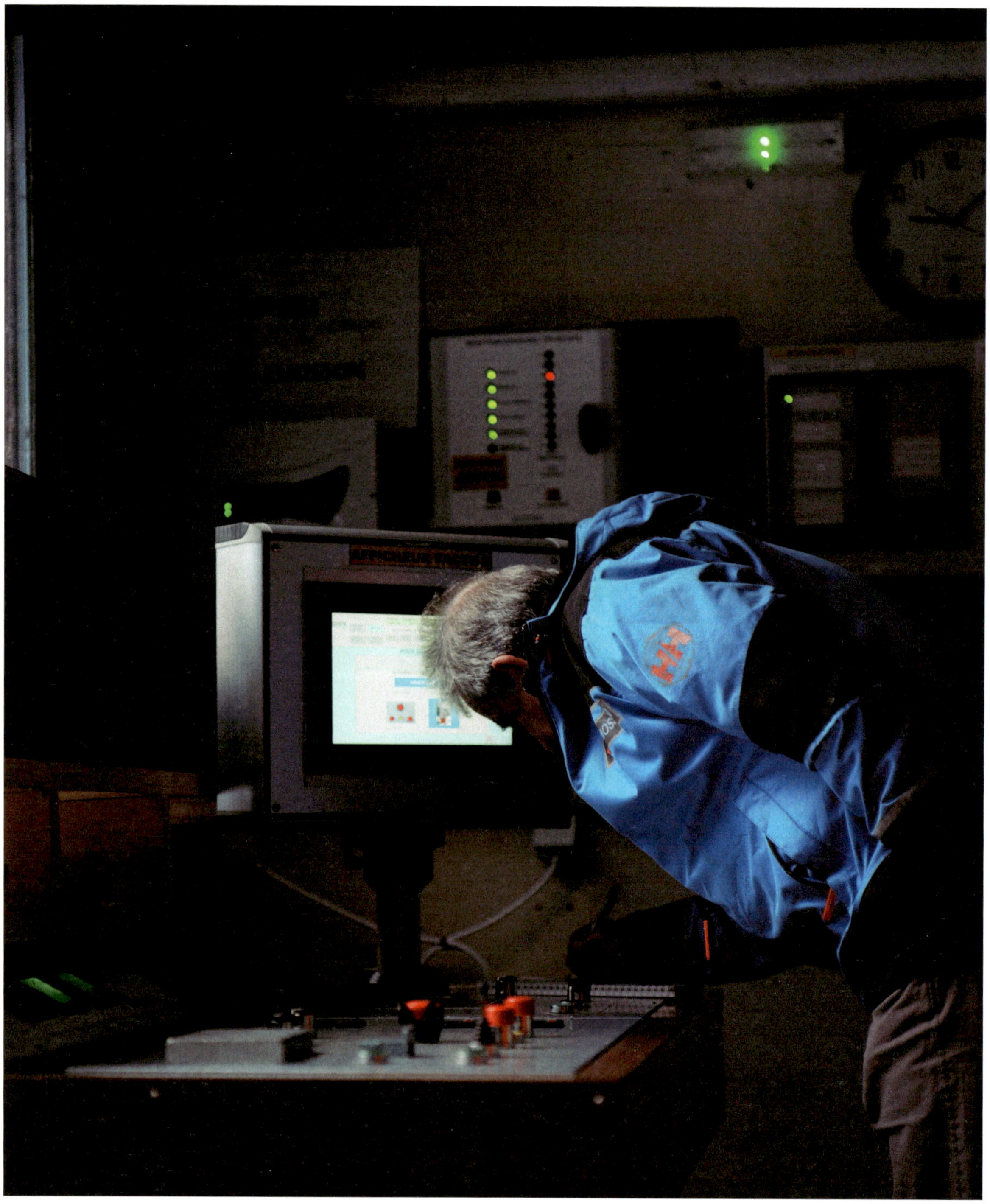

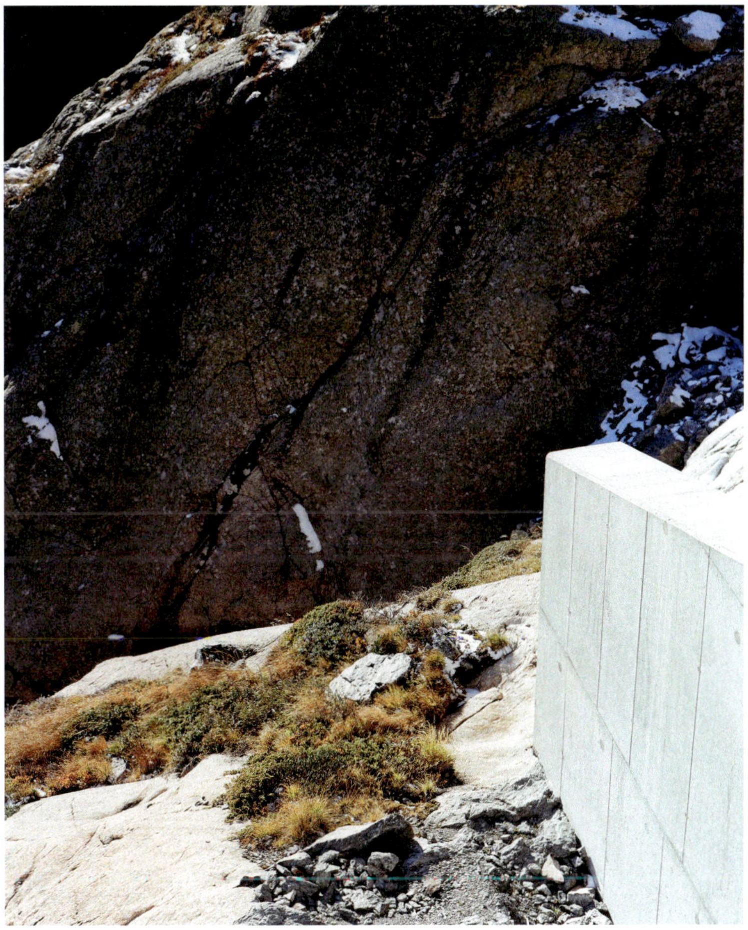

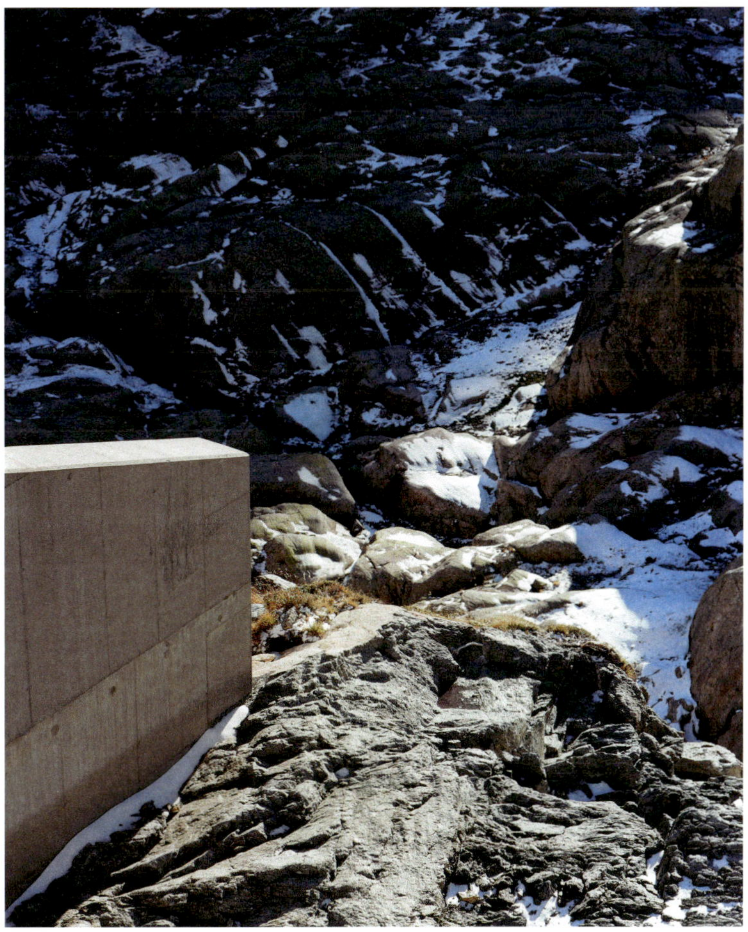

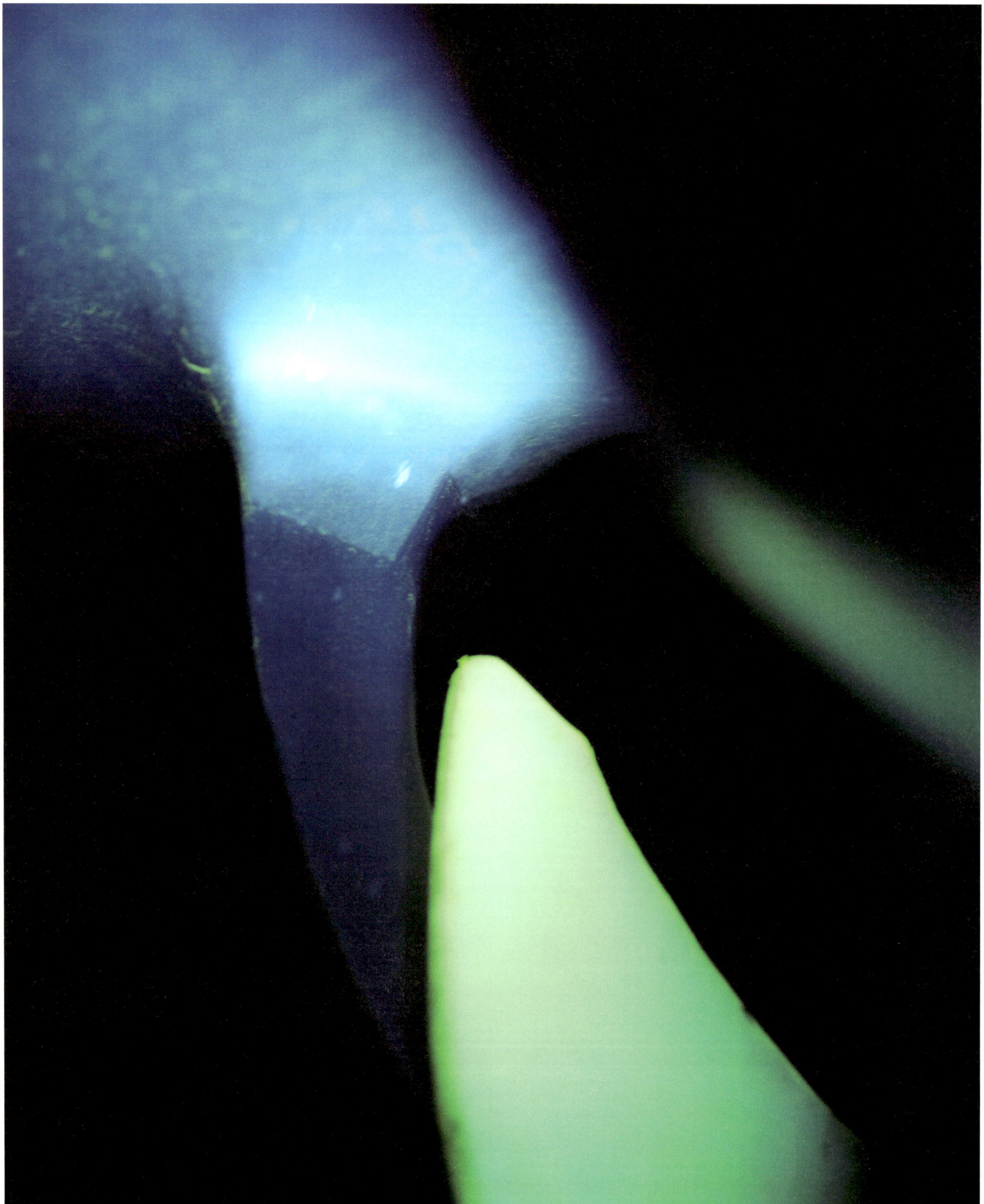

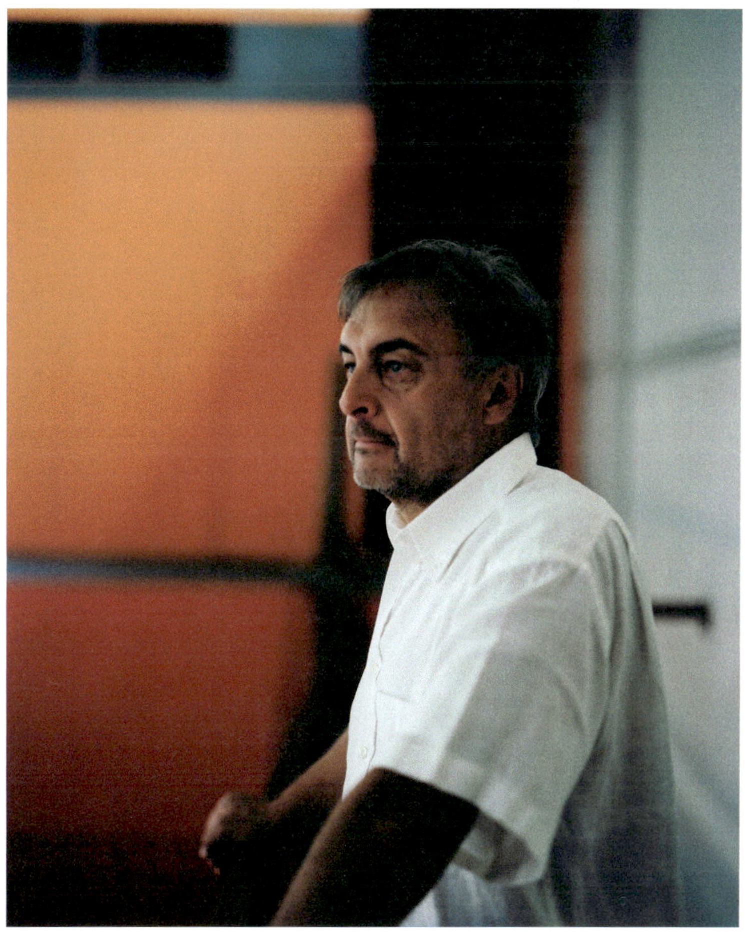

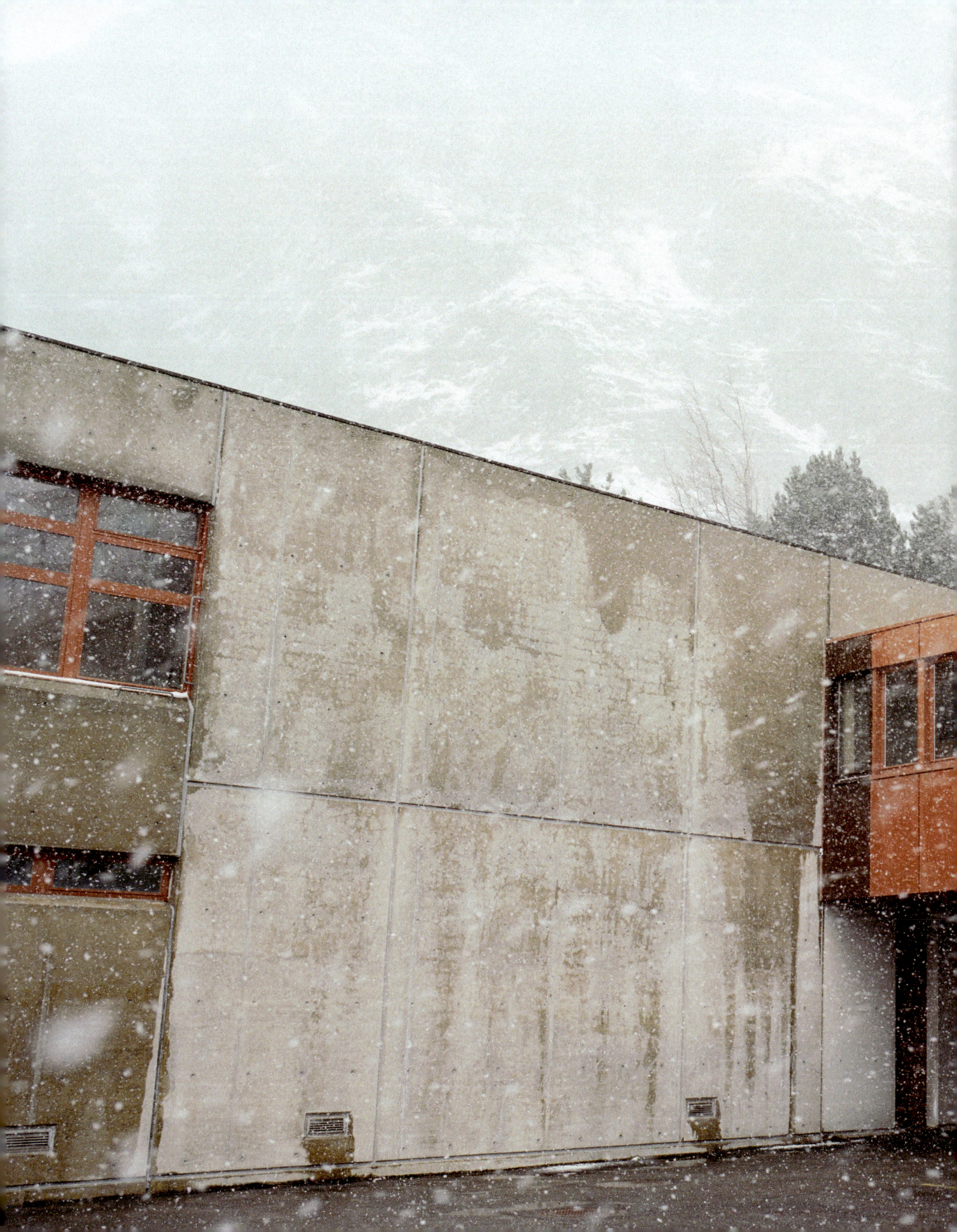

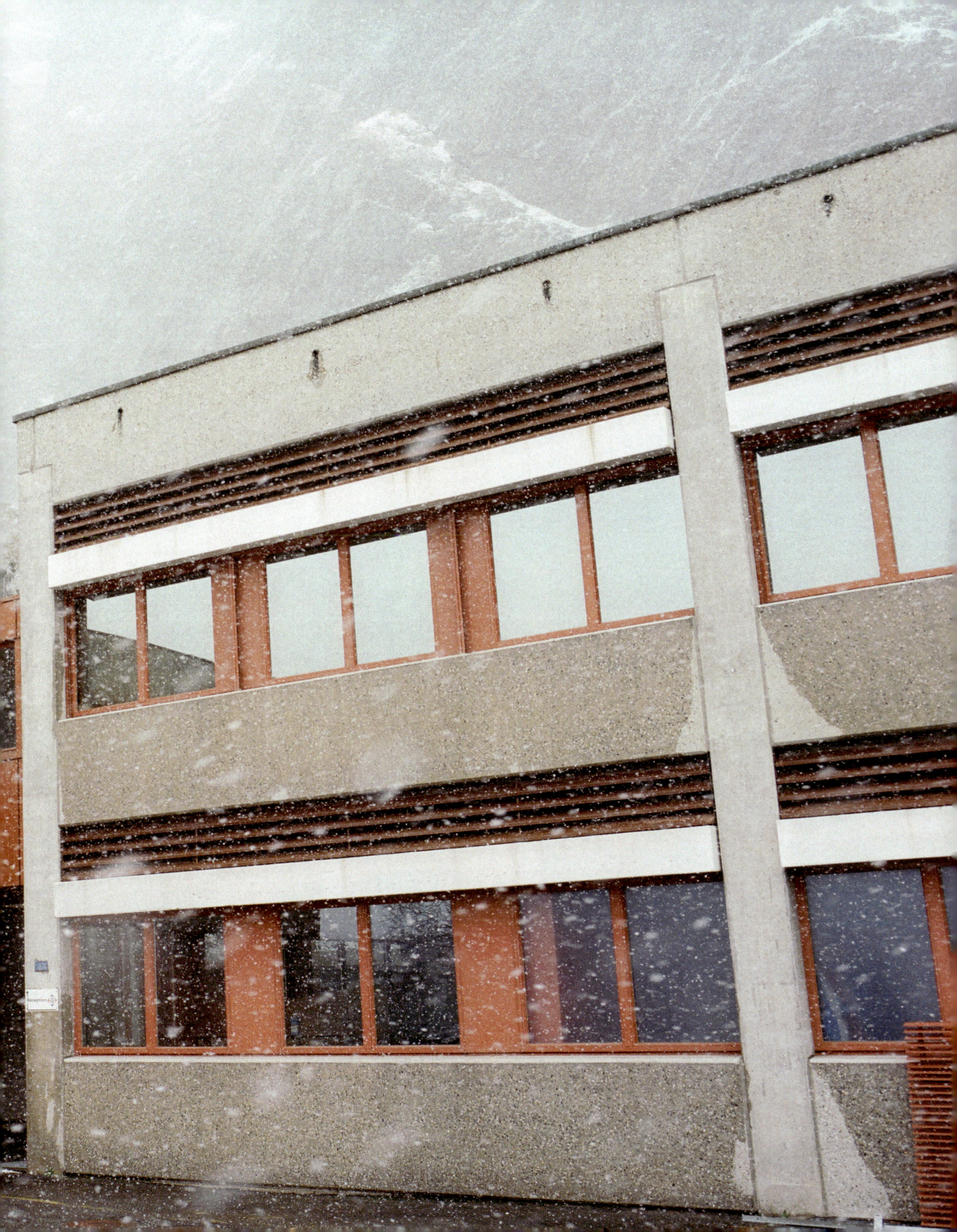
Reception

BK During our photographic expedition, we had the privilege of meeting Bruno Boulicaut, CEO of Electricité d'Emosson SA, meaning we could delve into the captivating world of hydropower. As an expert in this field, Mr. Boulicaut shared his knowledge about hydropower which stands at the forefront of electricity production, and highlighted its vital importance in everyday living.

BB As a child, like many kids of my age, I had fun playing with water, creating water mills and little turbines in small rivers and streams. My fascination for water may have unconsciously influenced my choice to work in the hydroelectricity sector. After engineering school, I had a great opportunity which allowed me to study hydraulic production and since 1992 I have dedicated my career to the hydroelectric field, first with Electricité de France (EDF) and then, from 2011, with Electricité d'Emosson SA. Throughout my career, I have therefore had the opportunity to explore several aspects of the energy sector, from electricity distribution to transportation.

Over the years, and within the dynamic context of managing hydroelectricity, our operations have been significantly influenced by external factors such as geopolitical tensions, energy supply, and also the COVID-19 pandemic. This has led to a deep transformation of energy markets, resulting in surging costs and instabilities within the supply network. The importance and dependability of our hydroelectric generation facilities have become indispensable for effectively navigating through these periods of unpredictability. The pandemic, in particular, has underscored the critical importance of energy security, driving us to ensure operational continuity while safeguarding the well-being of our workforce. In this context, hydroelectric power as a clean and renewable energy is a crucial factor in the global transition towards sustainable and carbon-neutral energy sources. This becomes particularly evident in nations such as Switzerland, where commitments to decarbonize energy systems and shift away from nuclear power are prominently emphasized. It fills me with pride to be part of an industry that holds such a significant role in addressing our energy needs, and it's truly meaningful to contribute to a future characterized by greater environmental awareness and sustainability.

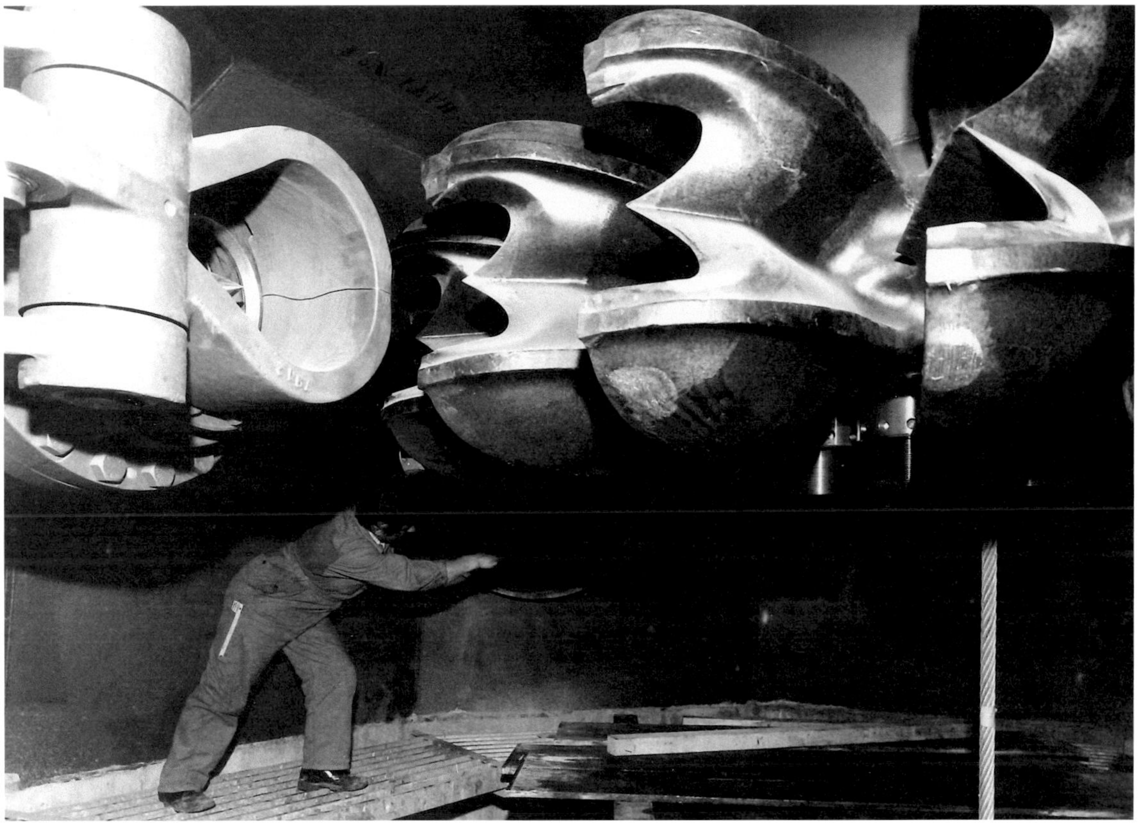

Bâtiaz Central, turbine,
Electricité d'Emosson SA,
M. Darbellay, 1972

The challenge of effectively storing electricity at a large scale requires aligning energy production with consumption to mitigate the potential risk of blackouts. The strength of hydroelectricity lies in its flexibility and reactivity, as we can store water in dams, such as the Emosson dam, to balance production and consumption. On the other hand, we must find the right balance between energy needs and the preservation of nature.

Looking to the future, we anticipate the transformation of glacial lakes into new sources for hydroelectric power generation. For instance, the Argentière Glacier is anticipated to undergo a transformation into a natural lake by approximately 2080. This shift represents a forthcoming opportunity for additional energy generation, requiring proactive planning to effectively harness this new potential.

This forward-thinking approach reflects our commitment to adapting to the changing environment while continuing to operate renewable energy resources. The impact of global warming is evident in our working environment, especially in terms of accelerated glacier melting and changes in water intakes in the Emosson dam. We are in the front line, closely monitoring the retreating glaciers and the increased water flow they bring. These circumstances are ambivalent, as they bring about both regrettable accelerated ice melting due to climate change, and simultaneously, an advantageous increase in available water for production.

Working in hydroelectricity is not only exciting but also rewarding; it allows me to stay connected with nature and the stunning mountains around. With unwavering dedication, we will continue to push the boundaries of innovation, finding new ways to maximize the potential of hydroelectric power while safeguarding our natural world. By doing so, we can look forward to a future powered by clean, renewable energy, and a planet that thrives alongside our ever-advancing civilization.

BK: Beatrice Gorelli and Keiichi Kitayama
BB: Bruno Boulicaut

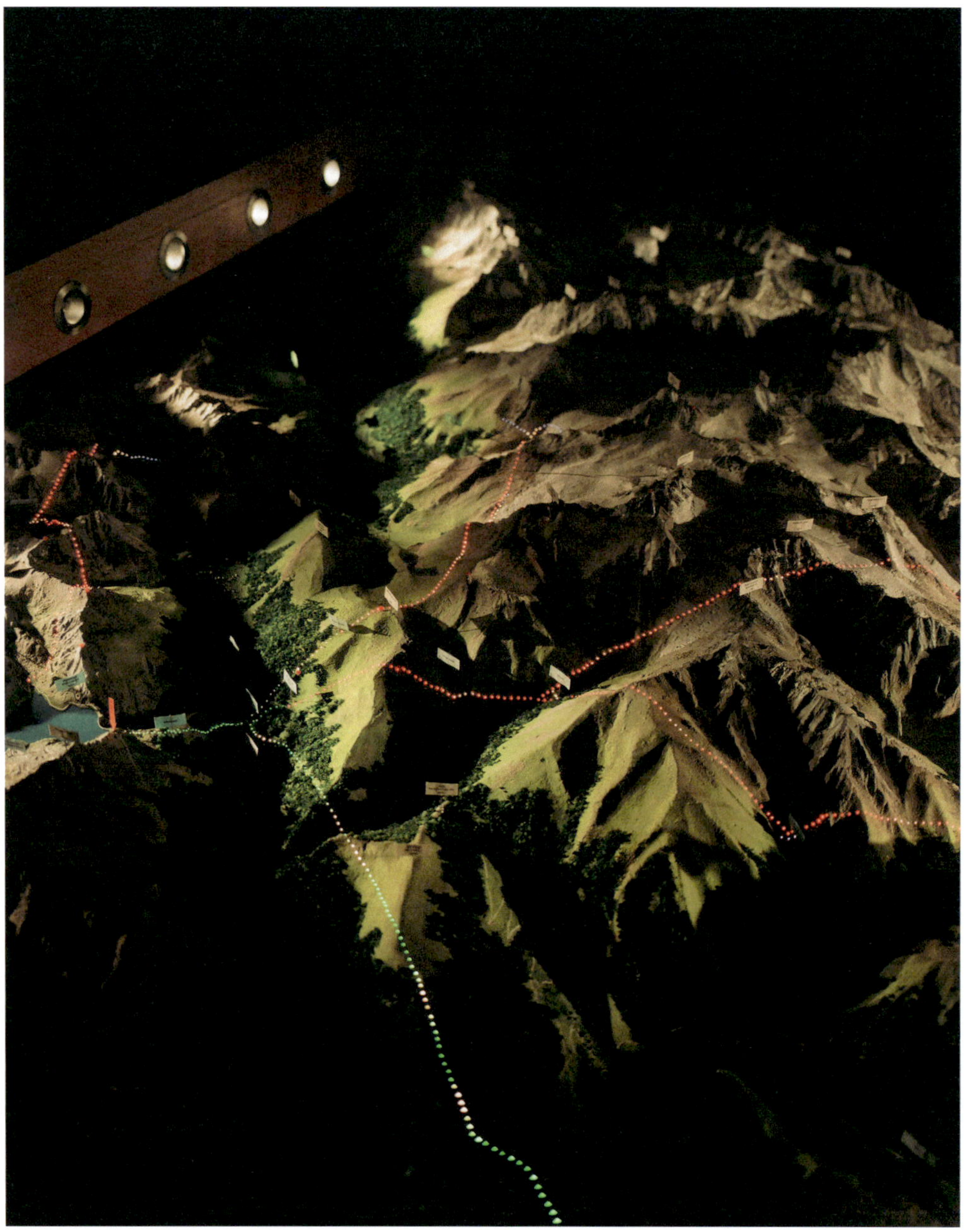

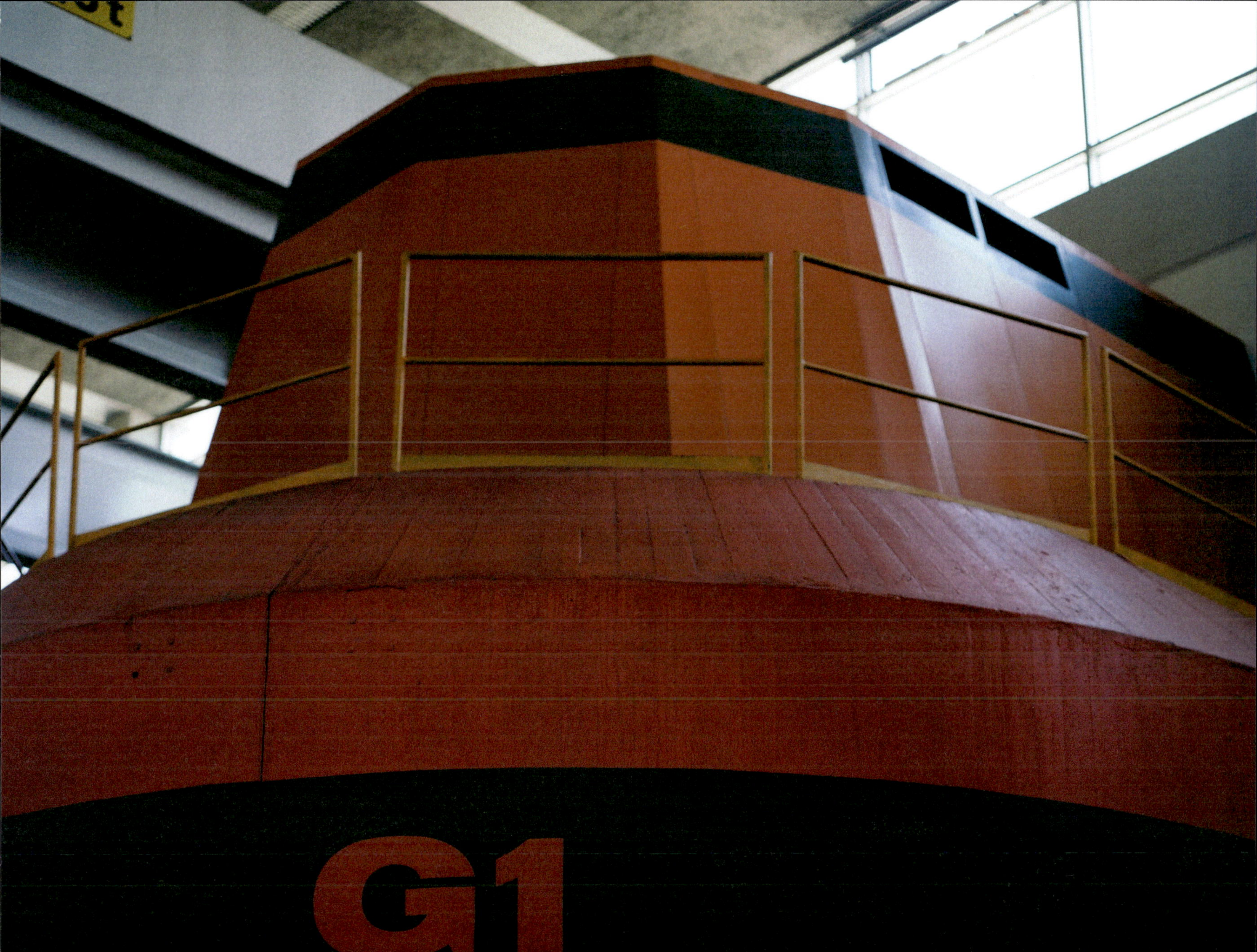
G1

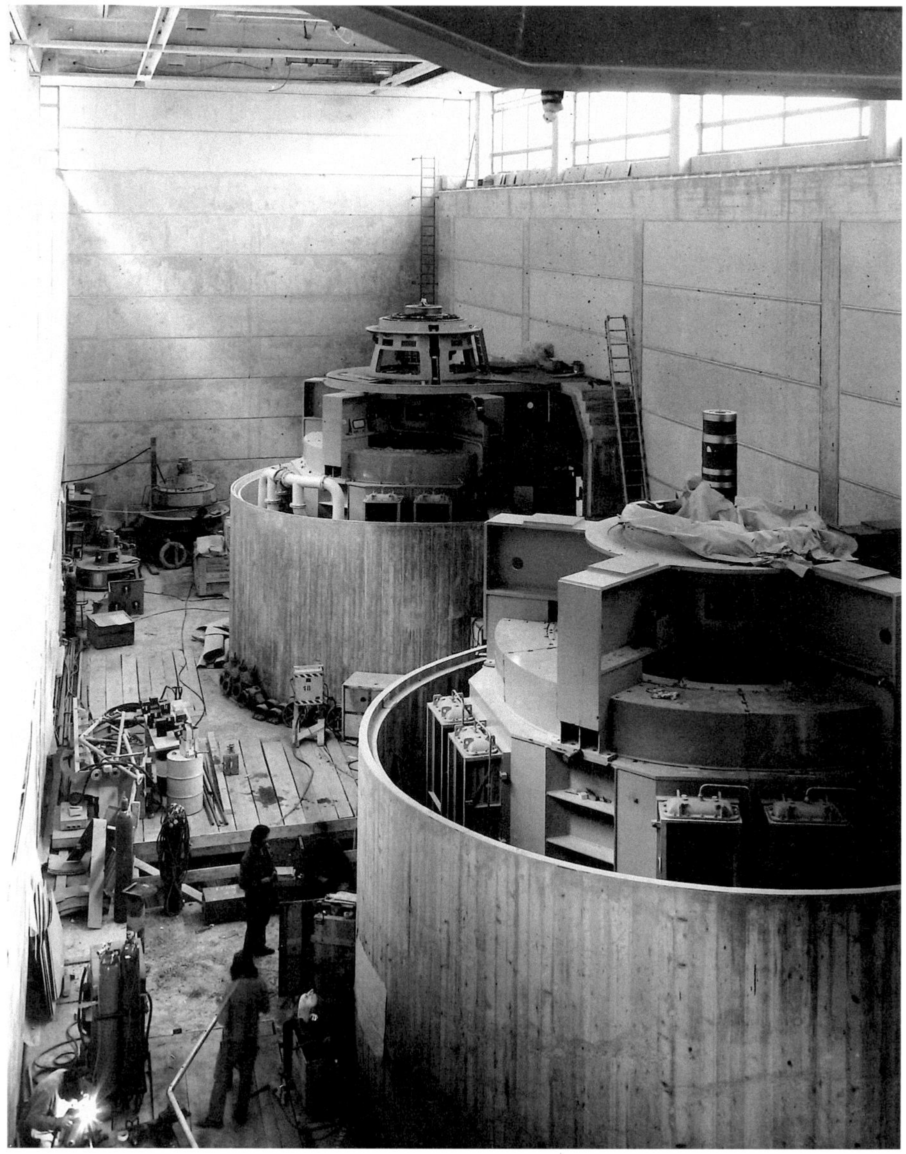

Bâtiaz Central, alternators, Electricité d'Emosson SA, M. Darbellay, 1972

BK　撮影をする中で、私たちはElectricité d'Emosson SA社のCEOであるブルーノ・ブリコー氏に会う機会を得ました。これにより、私たちは魅力的な水力発電の世界に触れることになります。水力発電分野の専門家であるブリコー氏は、電力生産の最前線に立つ水力発電に関する知識を共有し、我々の日常生活における重要性を説明してくれました。

BB　子供の頃、私は多くの同年代の子供たちと同様に、水で遊ぶことが楽しみで、小川に水車や小さなタービンを作って遊びました。水に魅了された経験は私が水力発電部門の仕事を選んだときに、無意識に影響したかもしれません。工業学校卒業後、私は水力発電に関して学ぶ機会を得て、1992年以来、私のキャリアを水力発電分野に捧げてきました。最初はフランス電力会社（EDF）で、その後2011年からはElectricité d'Emosson SA社で働いています。私はキャリアを通じて、電力の敷設から輸送まで、エネルギー部門のさまざまな側面を深く知る機会に恵まれました。

長年にわたり、水力発電管理の大局的な状況の中で、当社の事業運営は地政学的な緊張やエネルギー供給、そして新型コロナ感染症の大流行など様々な外部要因から大きな影響を受けてきました。これにより、エネルギー市場は大きく変革し、コストが急上昇し、供給ネットワークが不安定になりました。当社の水力発電設備の重要性と信頼性は、このような予測不能な時代を効果的に乗り越えるために不可欠となっています。特に新型コロナ感染症の大流行は、エネルギー安全保障の重要性を問い正し、作業員の健康を守りつつも、事業運営の継続性を確保する必要性を浮き彫りにしました。このような文脈において、

BK: Beatrice Gorelli and Keiichi Kitayama
BB: Bruno Boulicaut

水力発電はクリーンで再生可能なエネルギーとして、持続可能な脱炭素エネルギー源へのグローバルな移行において重要な要素となっています。これは特に、スイスなどの国々で顕著に強調される、エネルギーシステムの脱炭素化と原子力からの転換への取り組みにおいても明らかです。私はこのようなエネルギー需要に対処する上で、重要な役割を果たす業界の一員であることを誇りに思います。そして、環境意識と持続可能性が高まる未来に貢献することは、本当に意義深いことです。

大規模な電力を効果的に蓄積するという課題に対しては、停電の潜在的なリスクを軽減するために、電力生産を消費に合わせて適切に調整することが必要になります。水力発電の強みは、その柔軟性と即応力にあります。私たちはエモッソン・ダムが行なっているように、ダムに水を蓄えることで、生産と消費のバランスを取ることができます。一方で、エネルギーの需要と自然保護との間においても、適切なバランスを見つけなければなりません。

将来的には、氷河湖が水力発電の新たな供給源になることが予想されています。例えば、アルジャンティエール氷河は、およそ2080年までに自然湖になることが予測されています。この変化は、新たなエネルギー生産の可能性を示しており、この新たな可能性を効果的に活用するためには先を見越した計画が必要です。

この未来を見据えたアプローチは、再生可能エネルギー資源を利用し続けながら、変化する環境に適応するという当社の取り組みを反映しています。地球温暖化の影響は、特に氷河の急速な融解や、エモッソン・ダムへの流入水量の変化など、当社の作業環境において顕著に見られます。当社は最前線で、氷河の後退とそれに伴う水量の増加を注意深く監視しています。現在の状況は、相反する二つの側面を持っています。気候変動による氷河の融解の加速という嘆かわしい事態をもたらすと同時に、エネルギー生産に利用可能な水の増加という有利な事態をもたらすからです。

水力発電に関する仕事は、刺激的なだけでなく、やりがいもあります。自然や周囲の美しい山々との繋がりを保つことができるのですから。揺るぎない献身をもって、私たちはイノベーションの限界に挑み、水力発電の可能性を最大限に引き出しながら、自然環境を守ります。これによって、クリーンかつ再生可能なエネルギーを動力源とする未来と、進歩し続ける文明と共に繁栄する地球を待ち望むことができるのです。

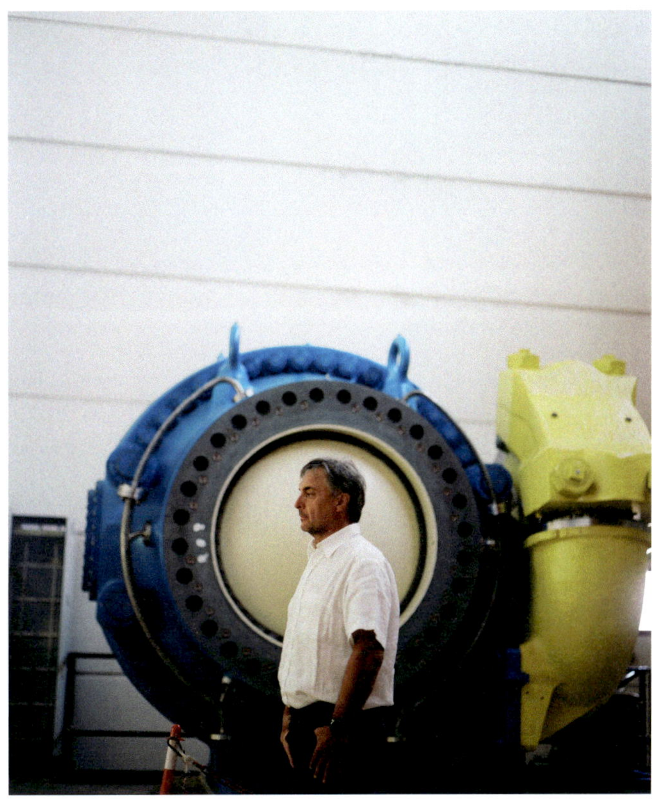

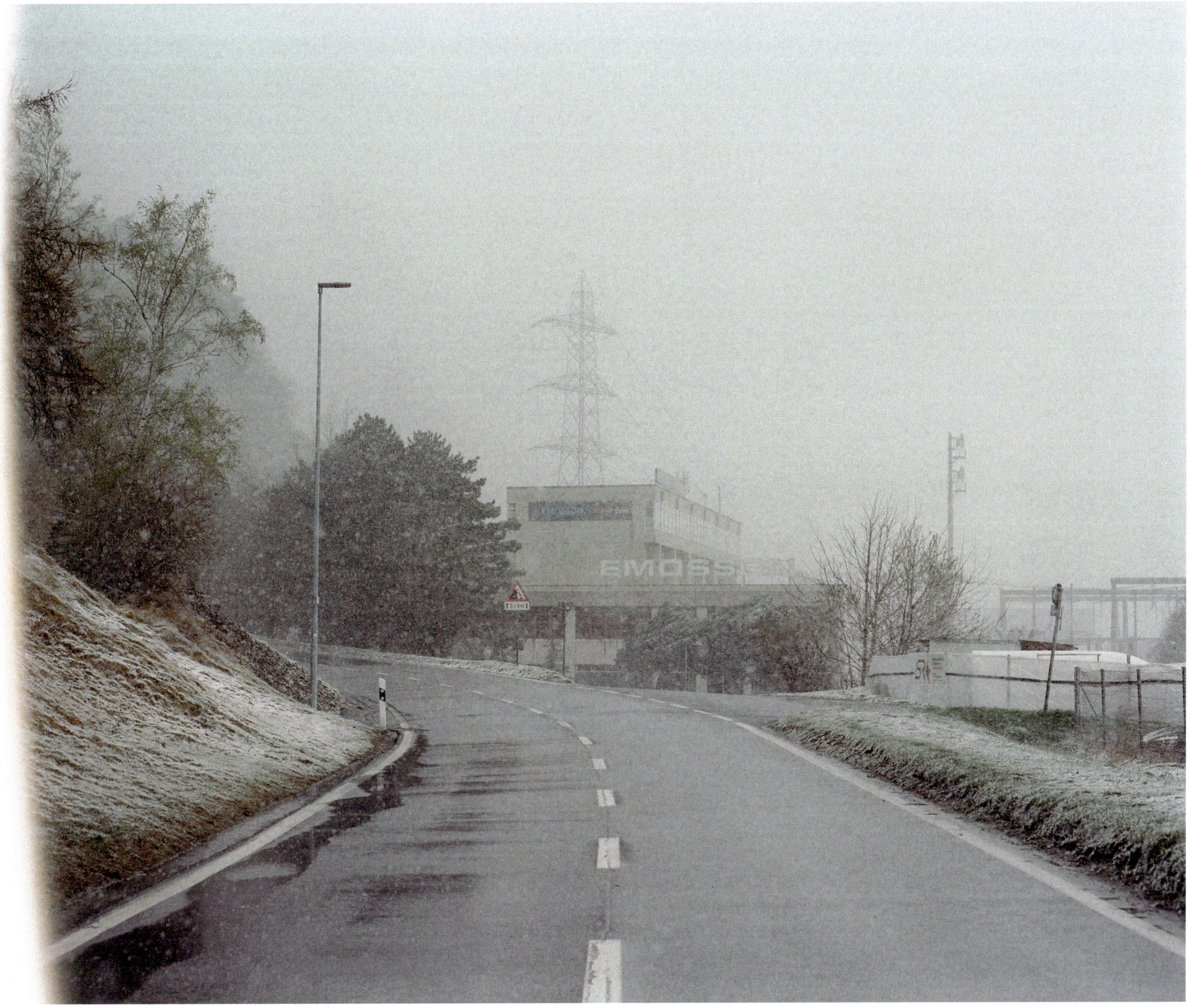
EMOSS

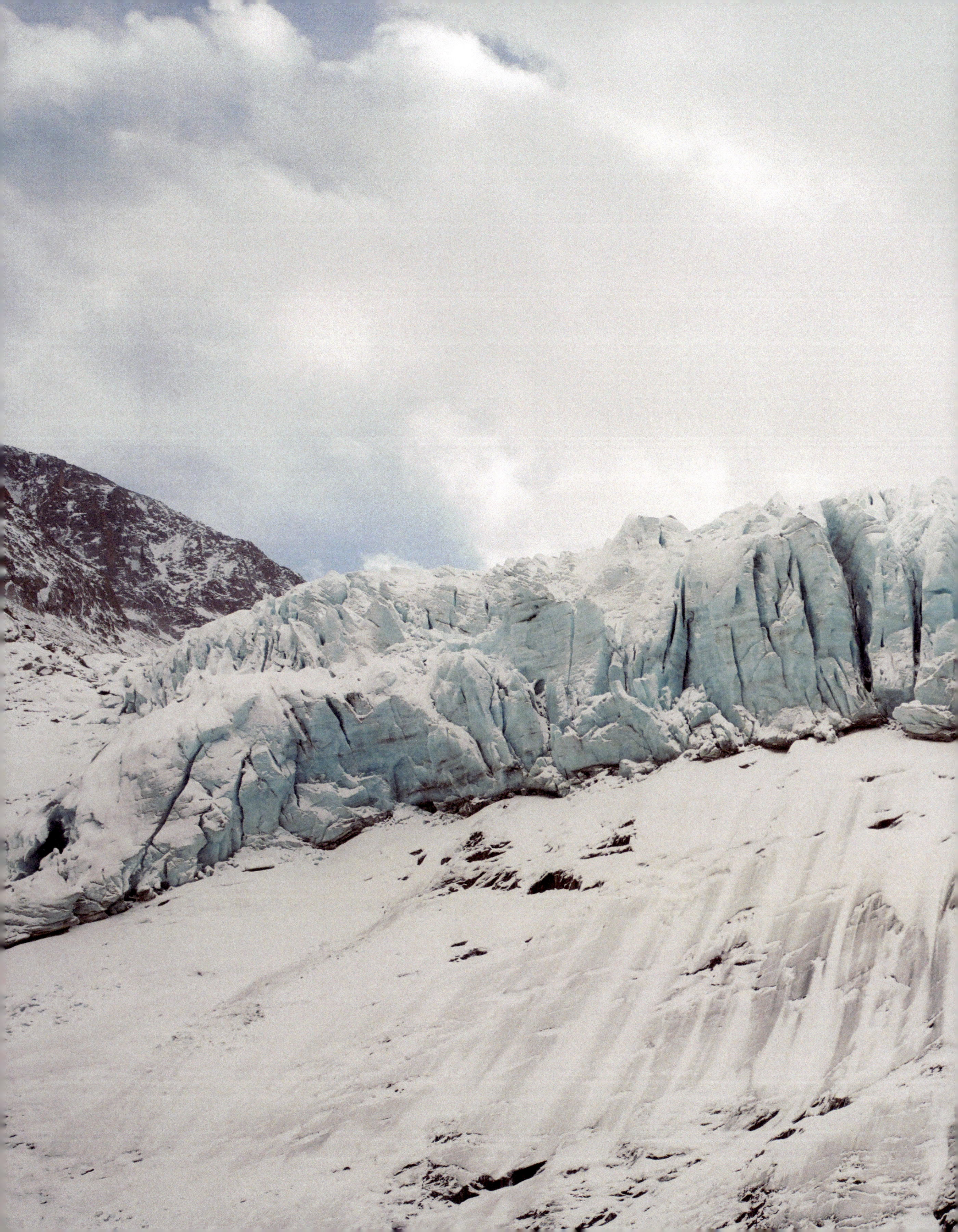

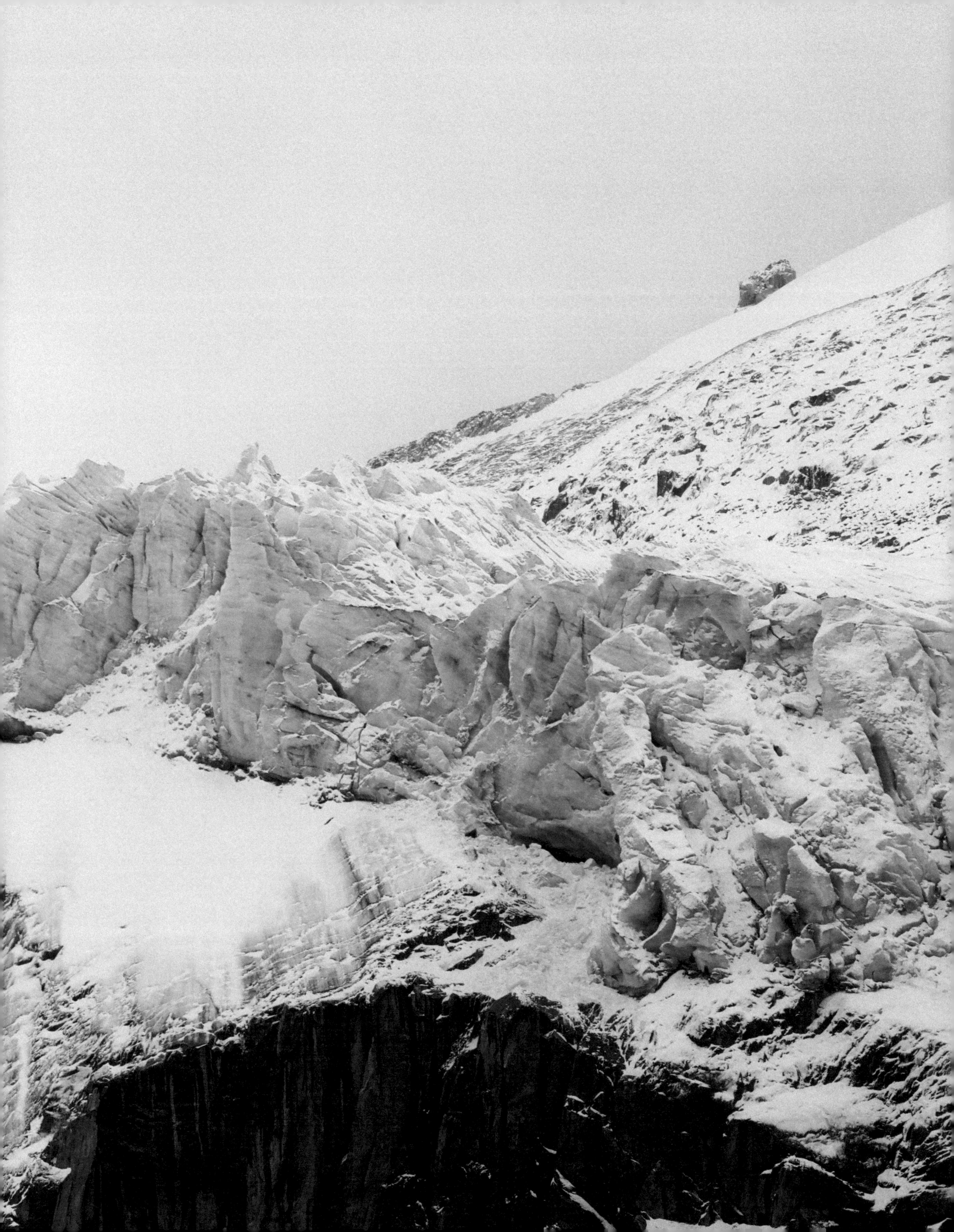

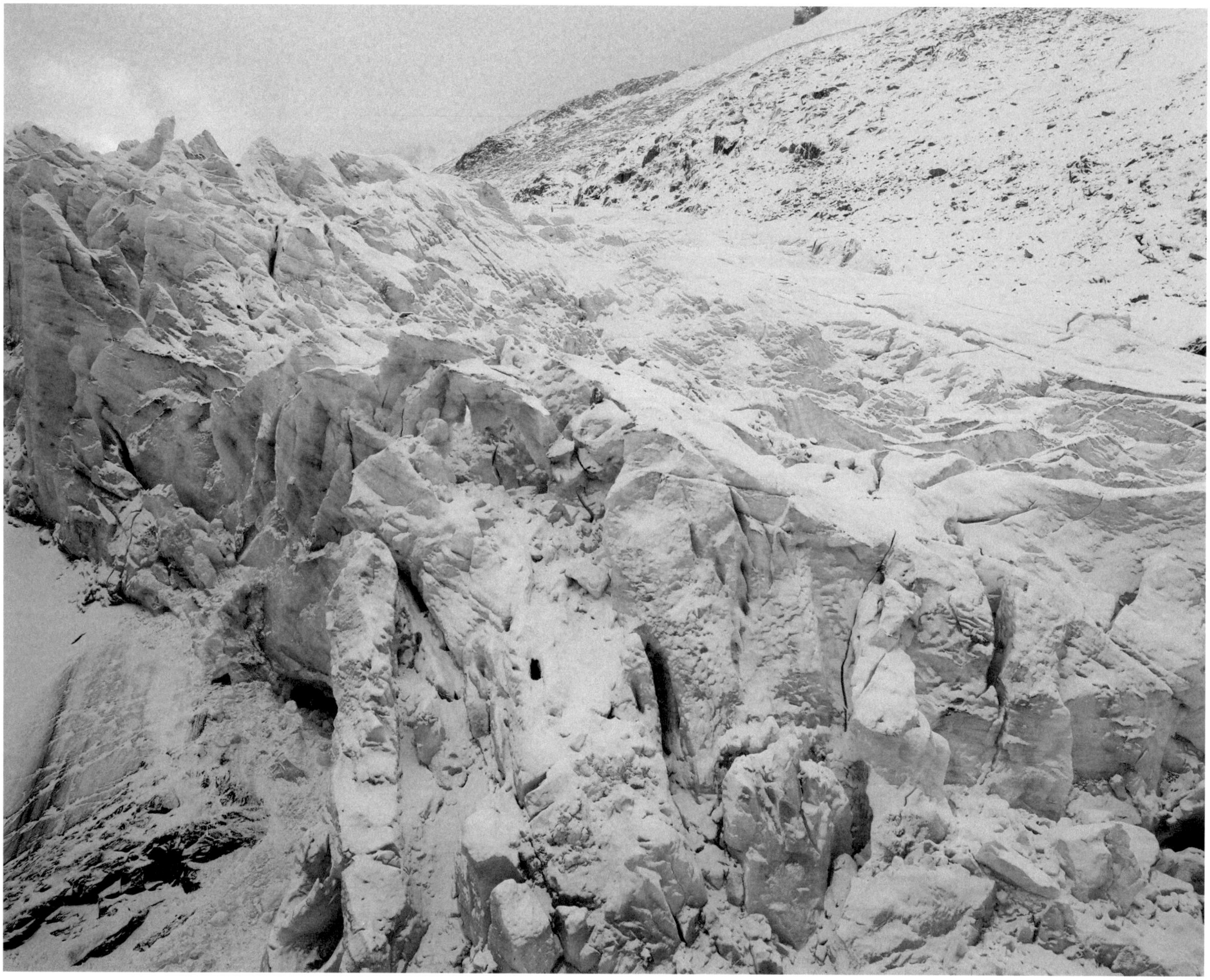

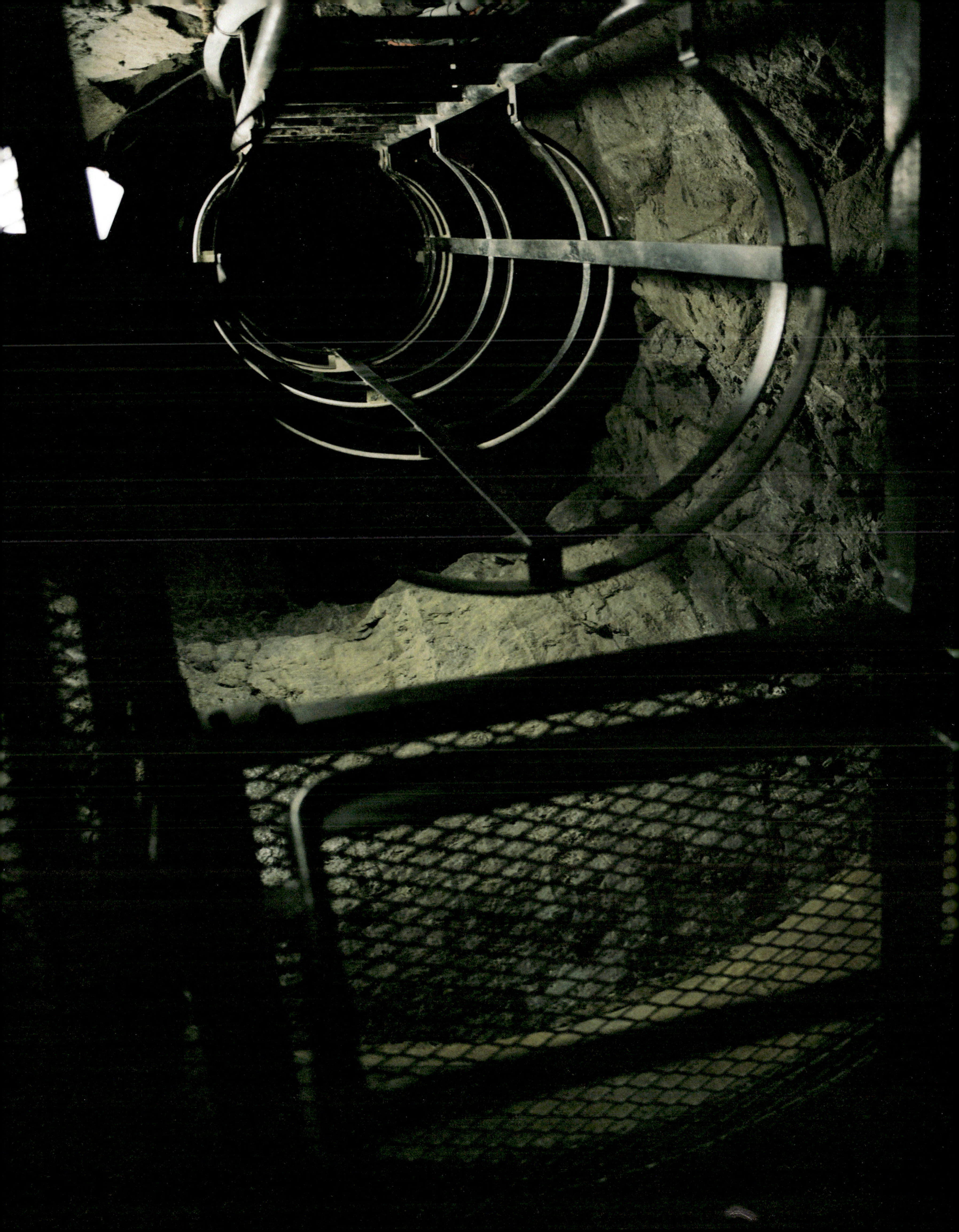

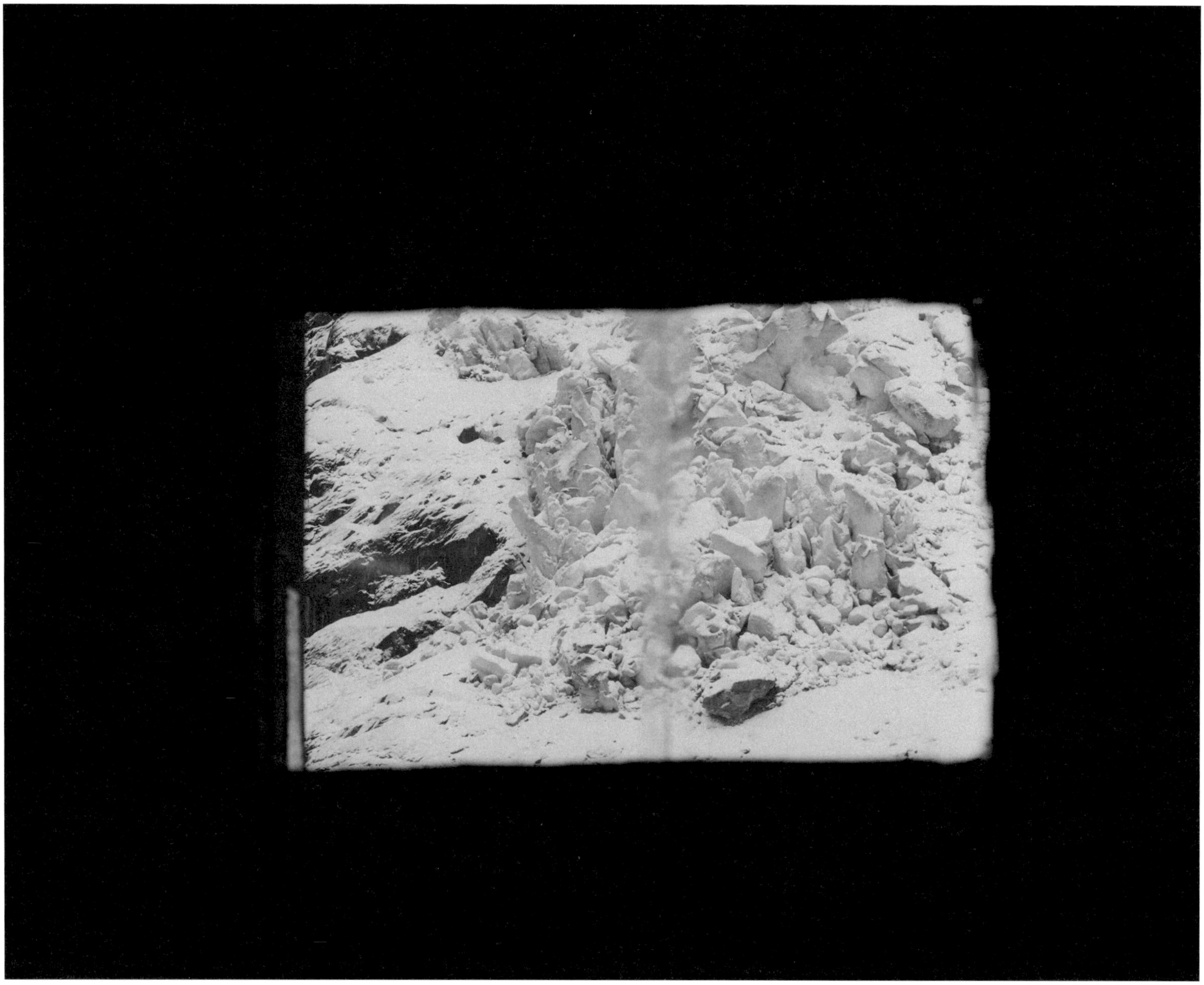

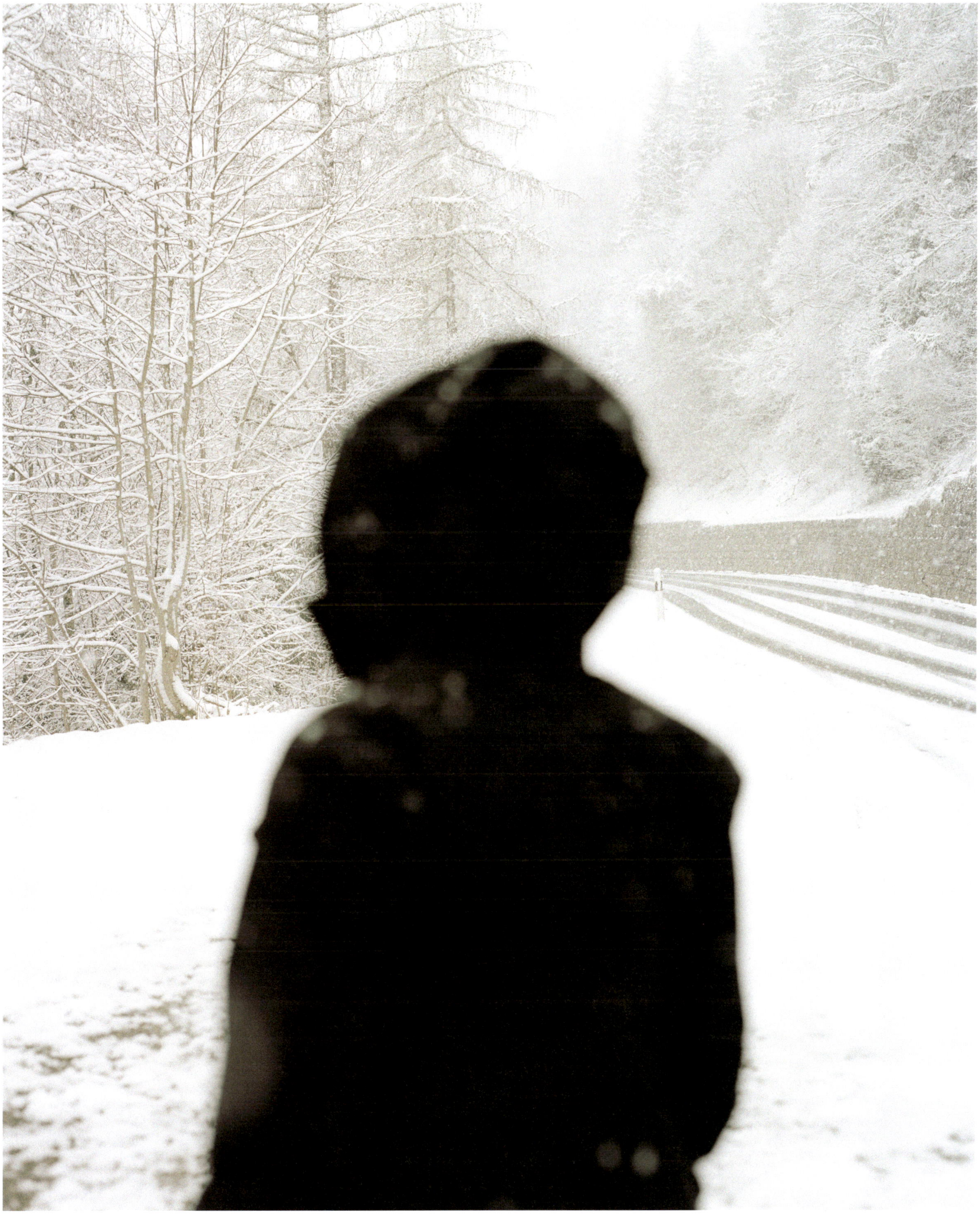

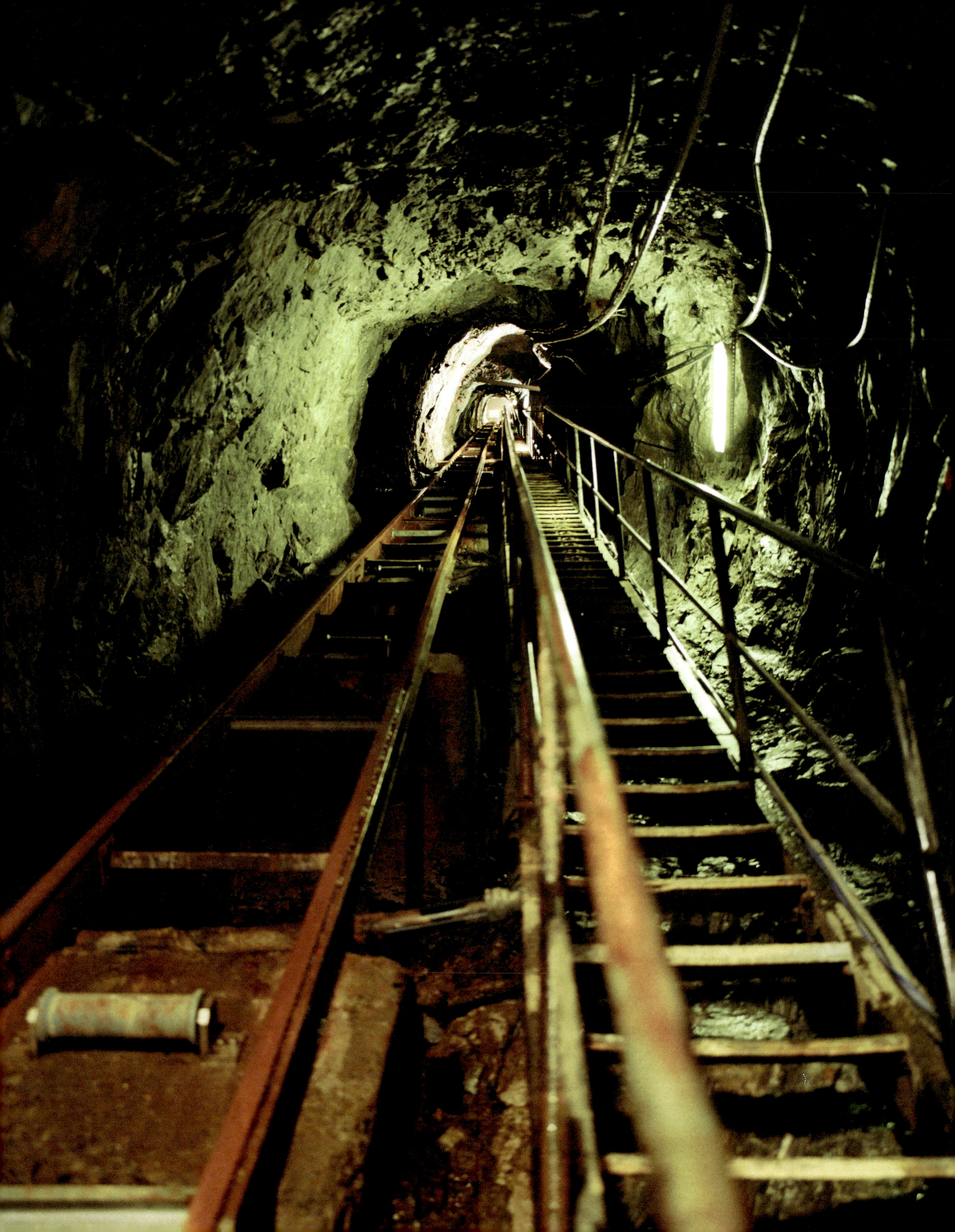

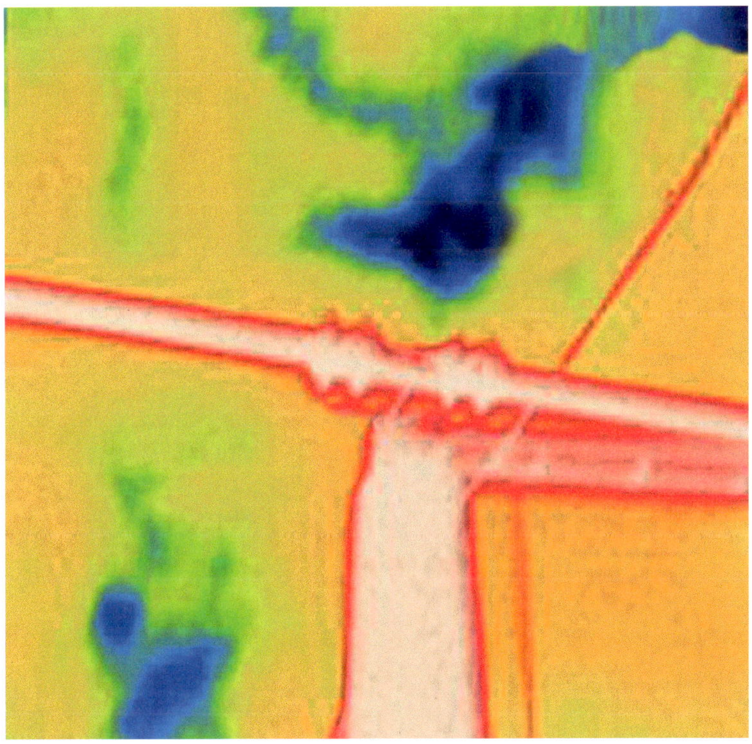

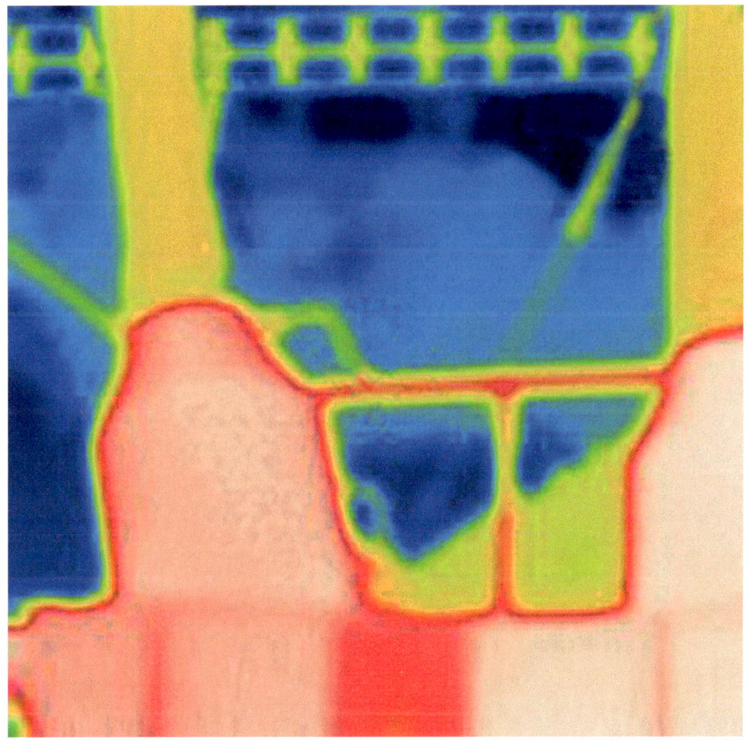

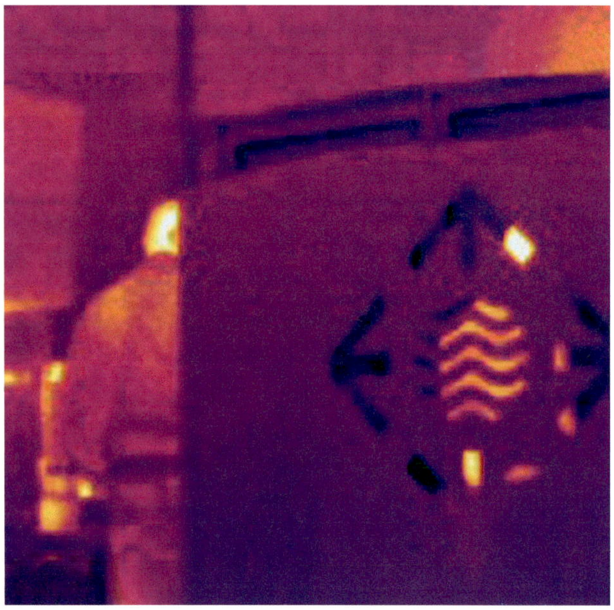

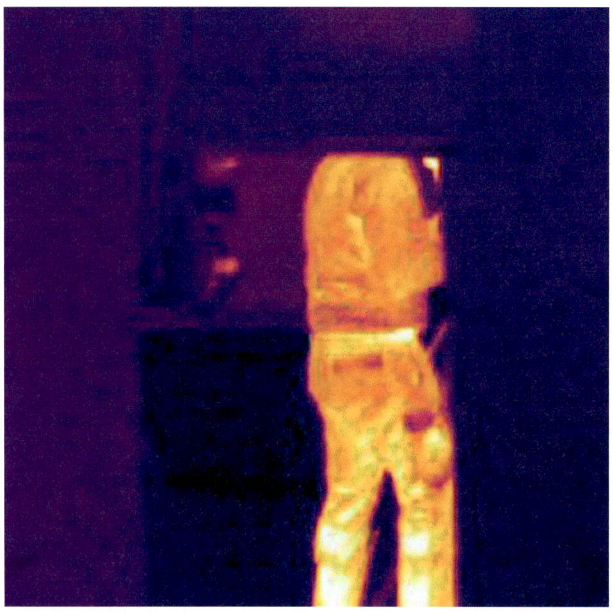

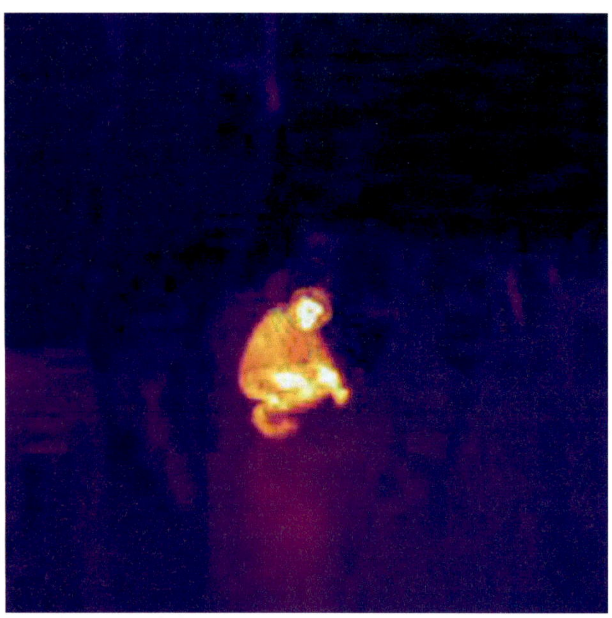

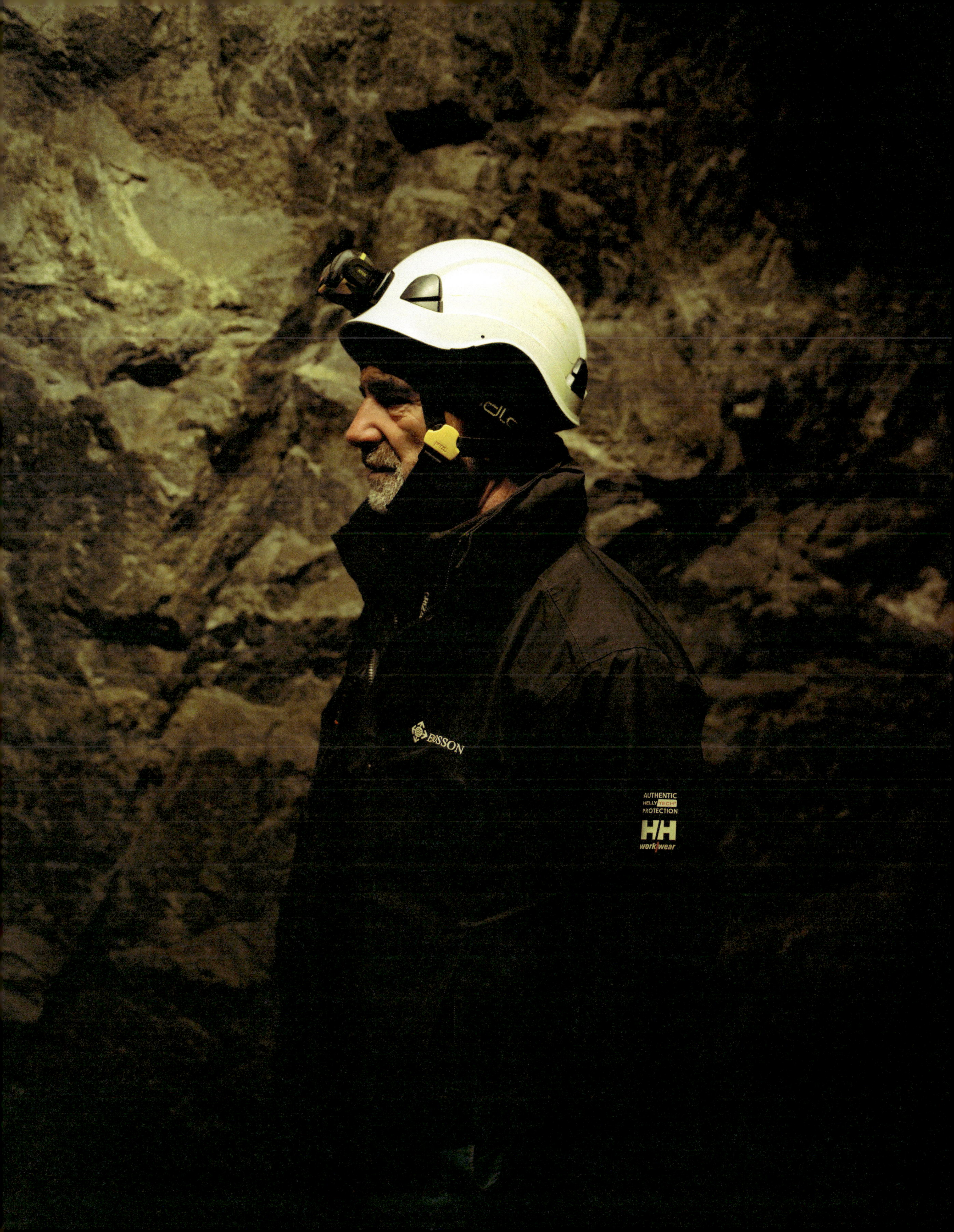
AUTHENTIC
HELLY TECH
PROTECTION
HH
workwear

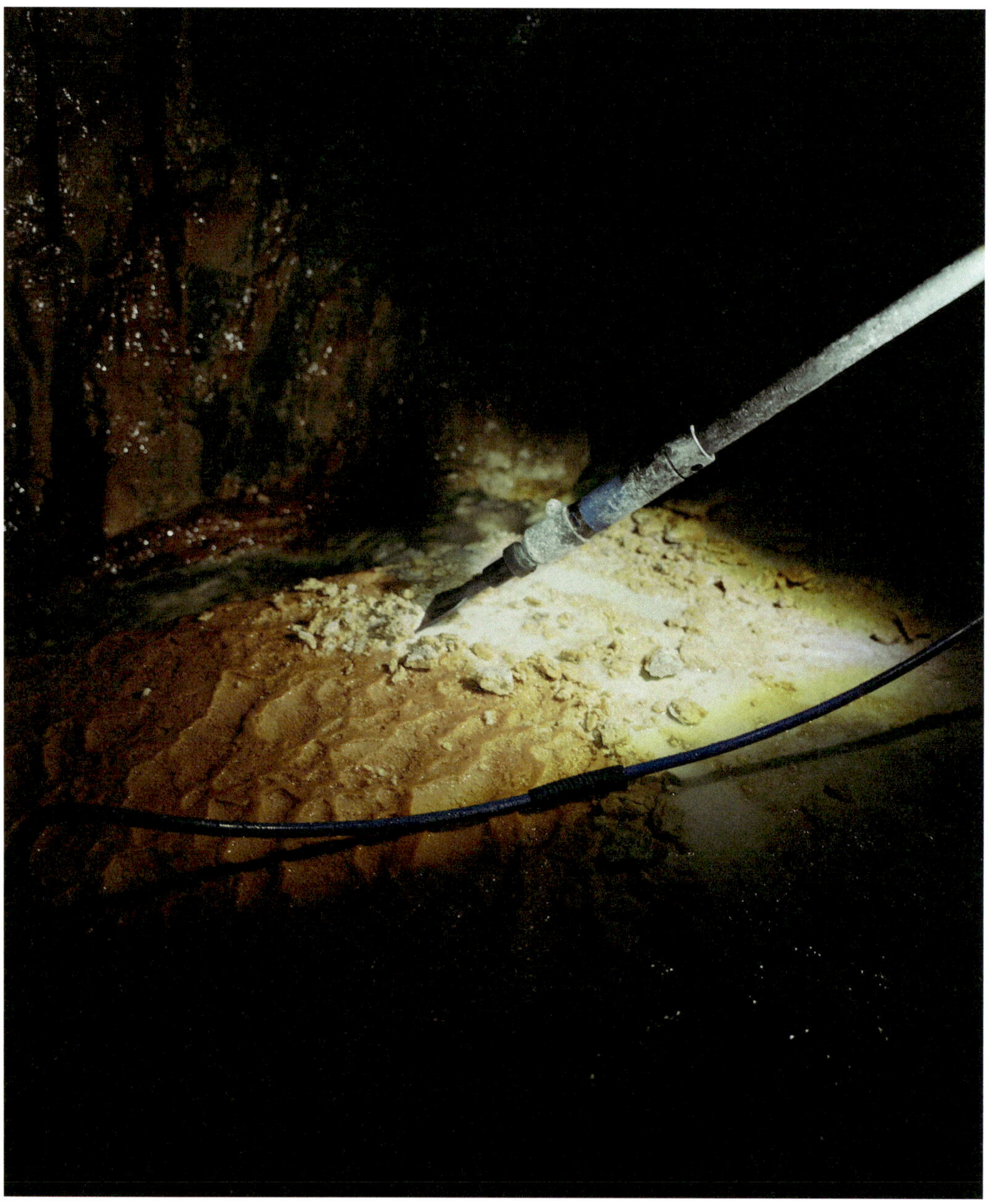

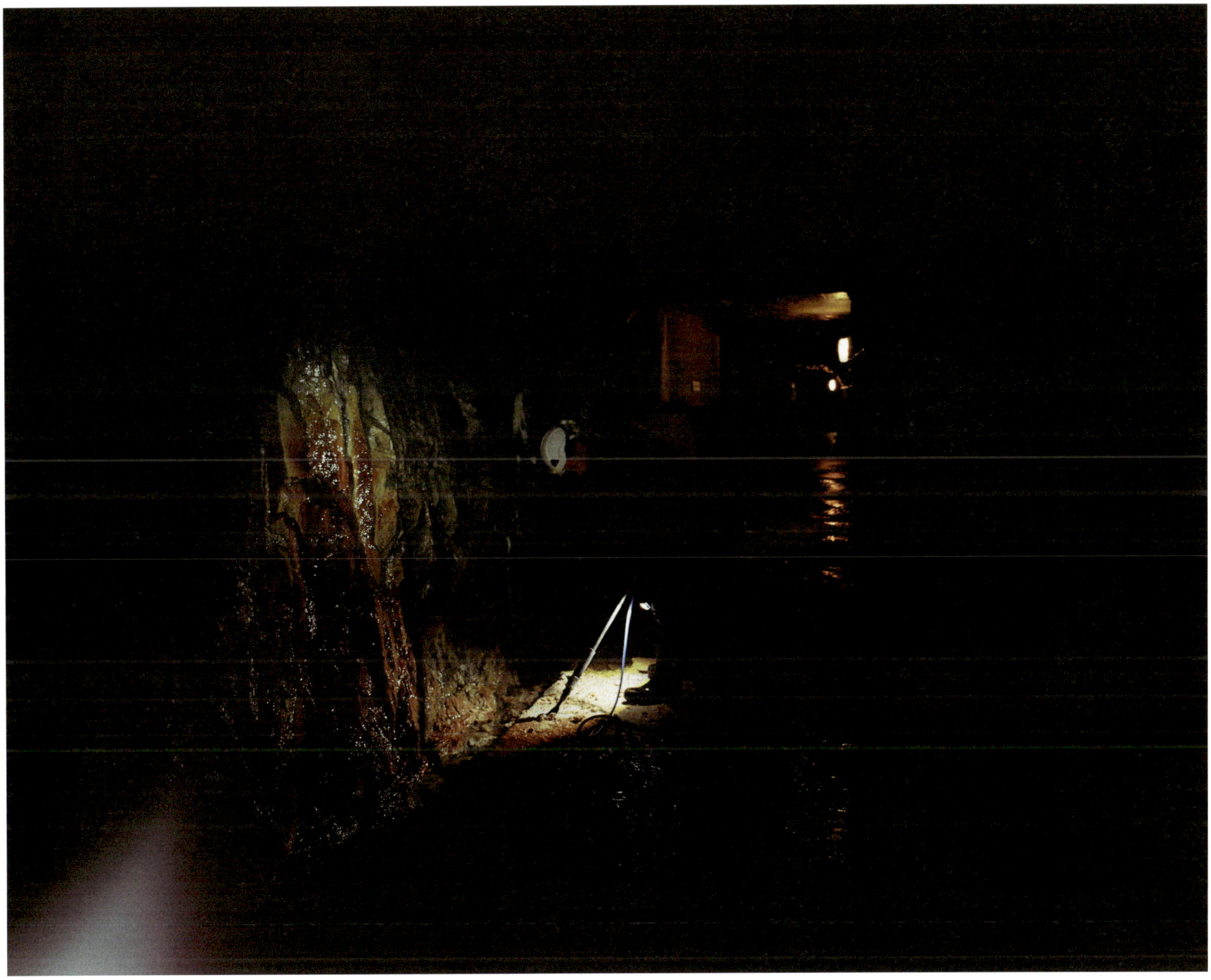

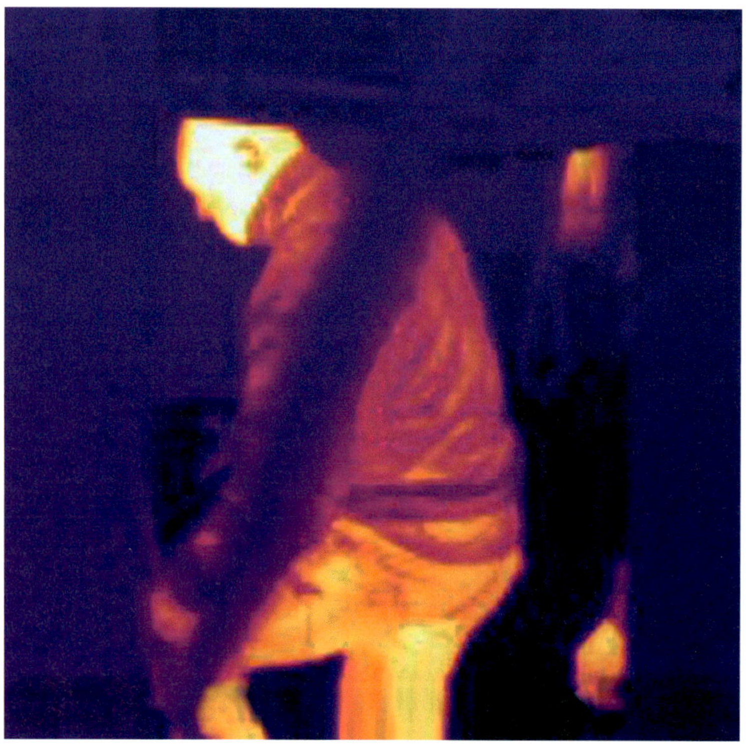

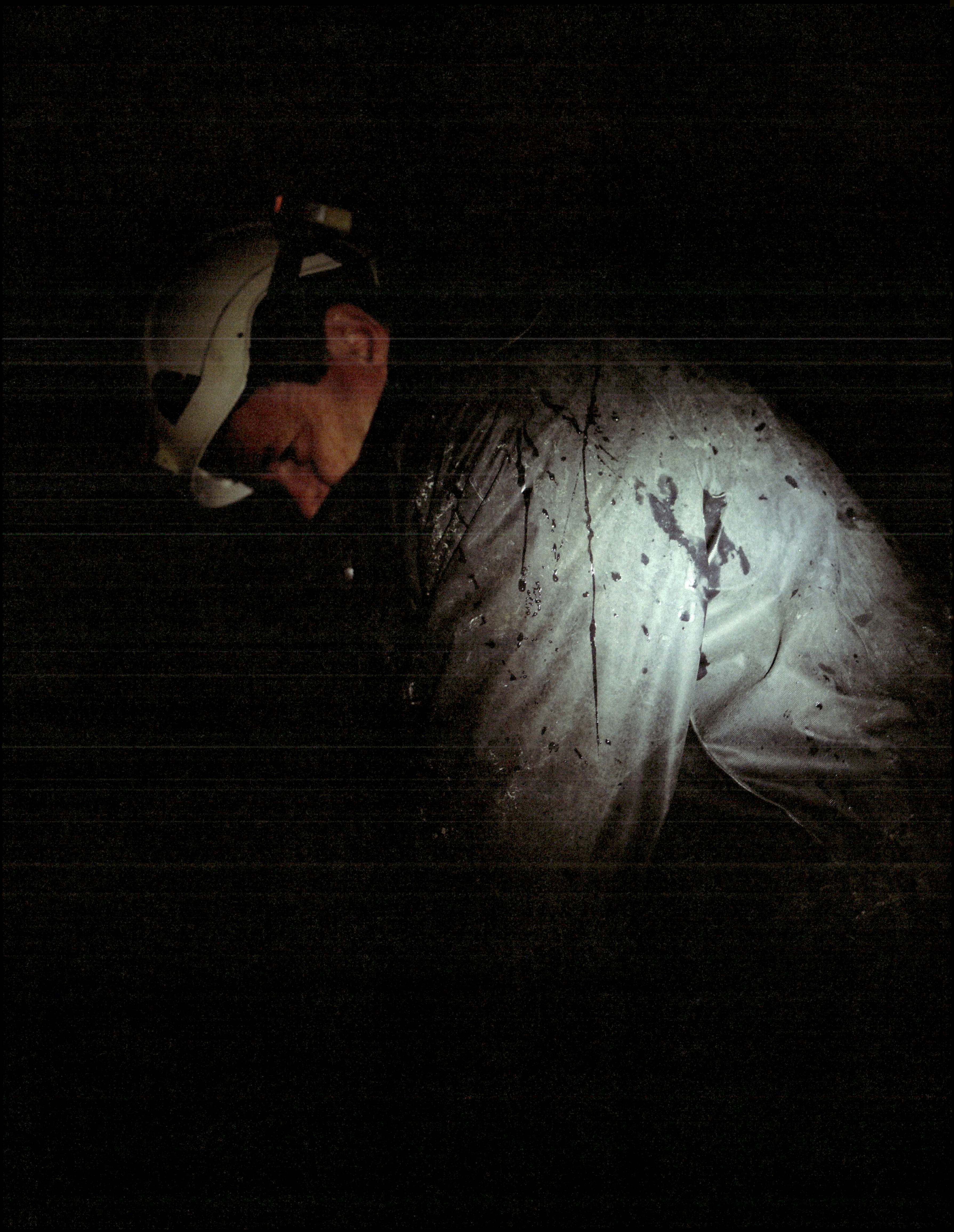

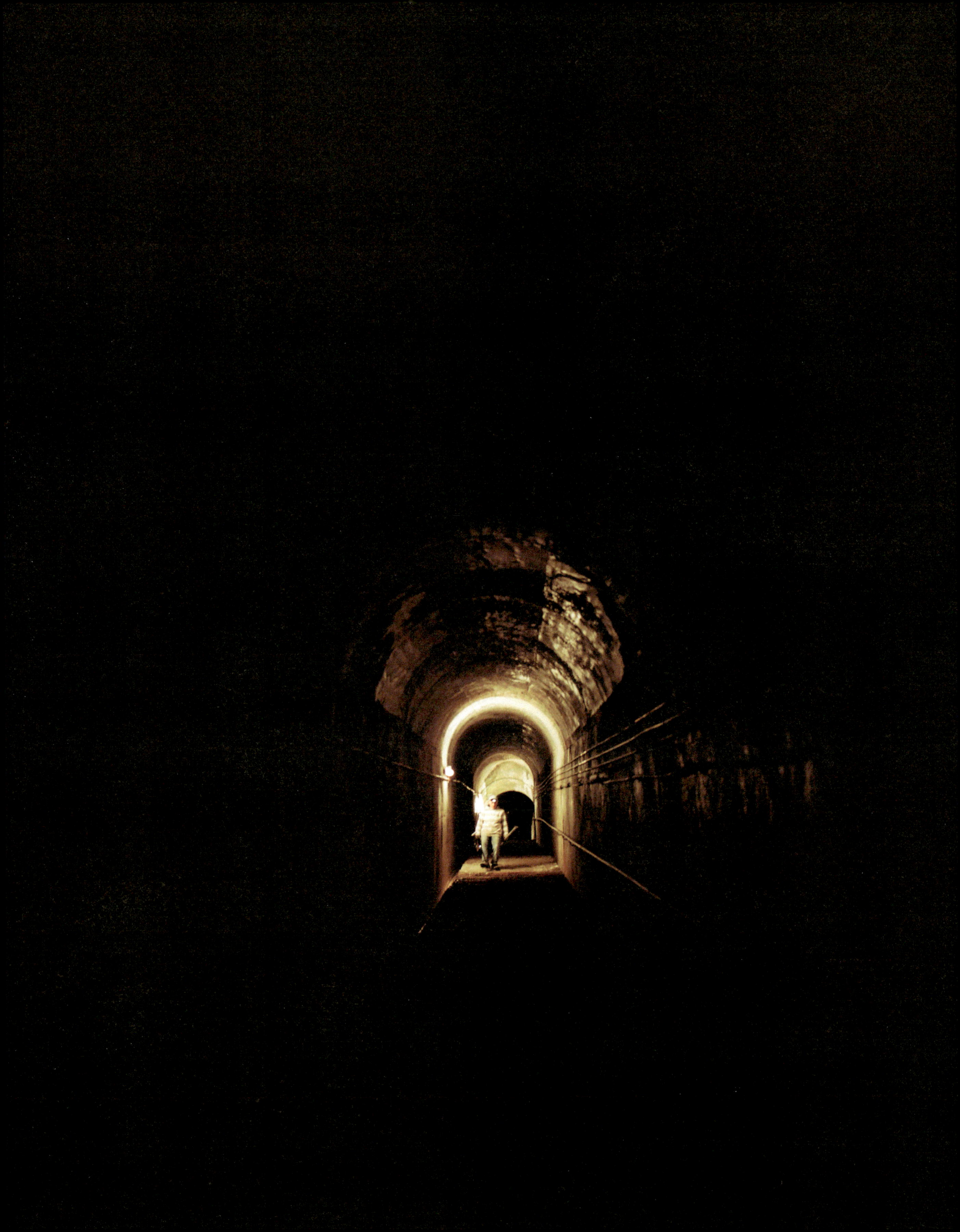

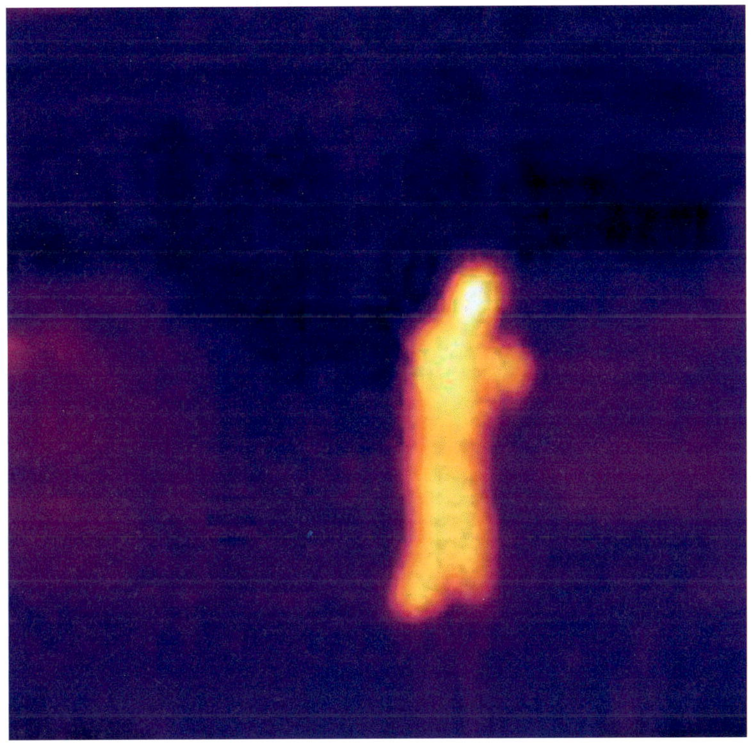

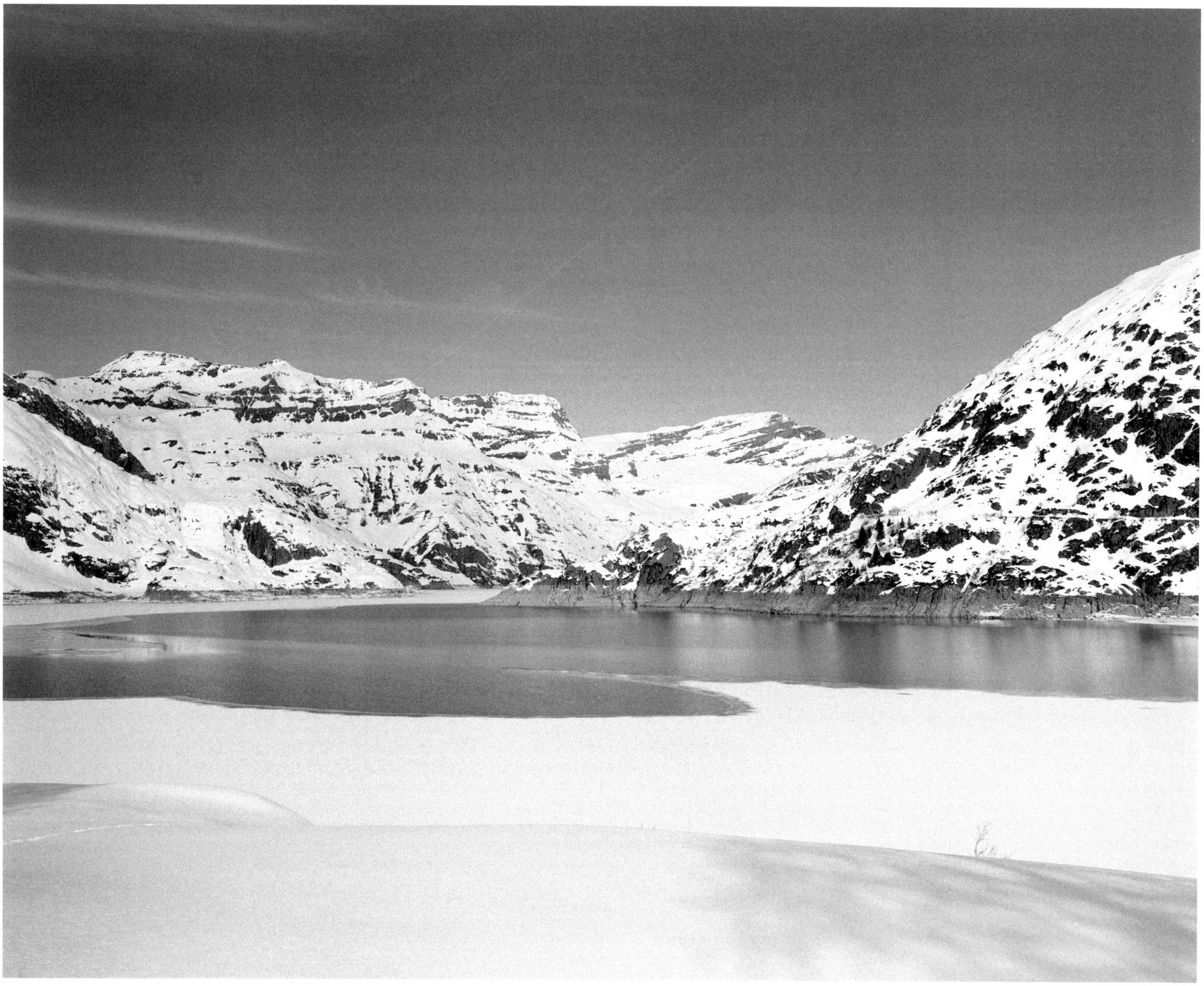

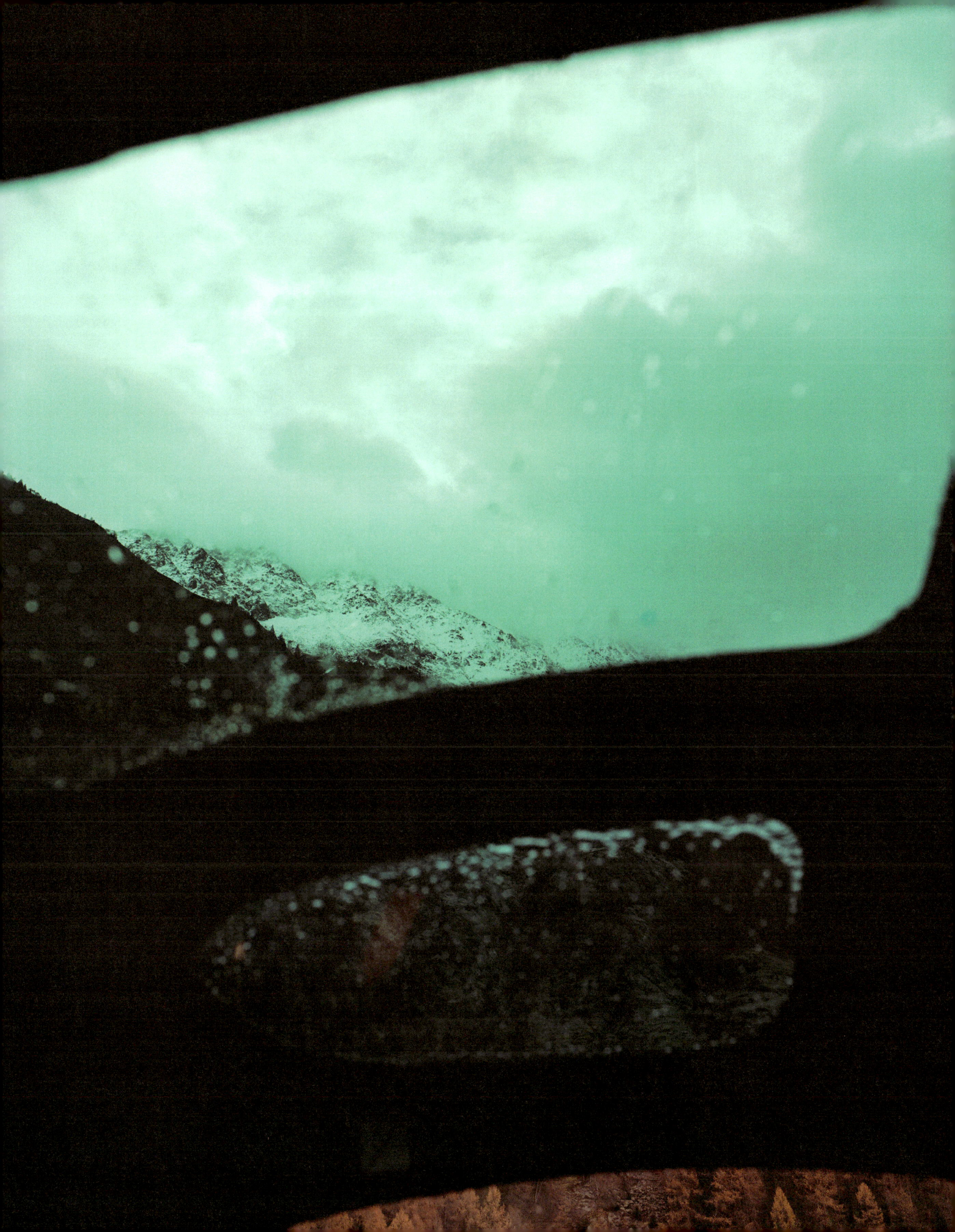

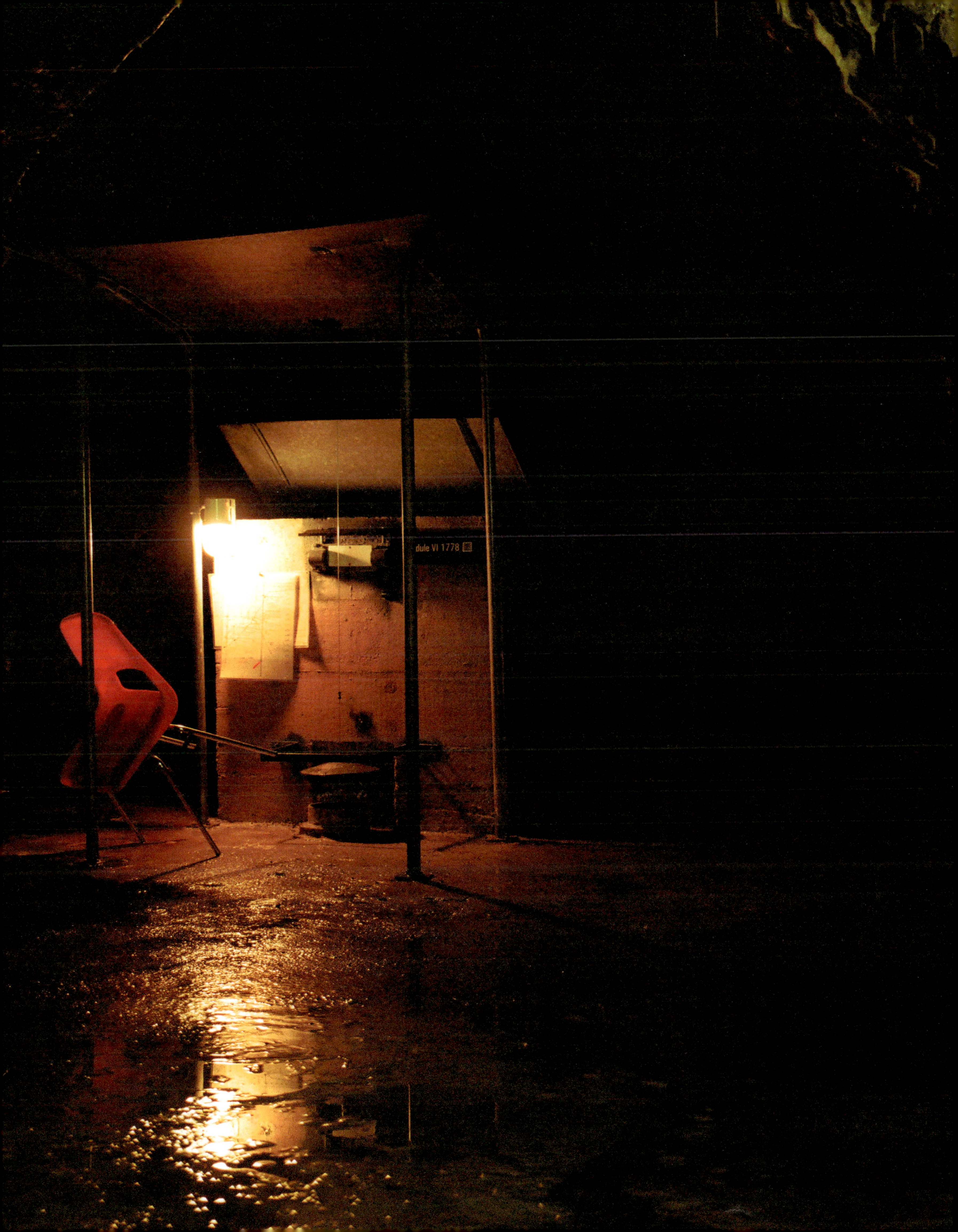
dule VI 1778

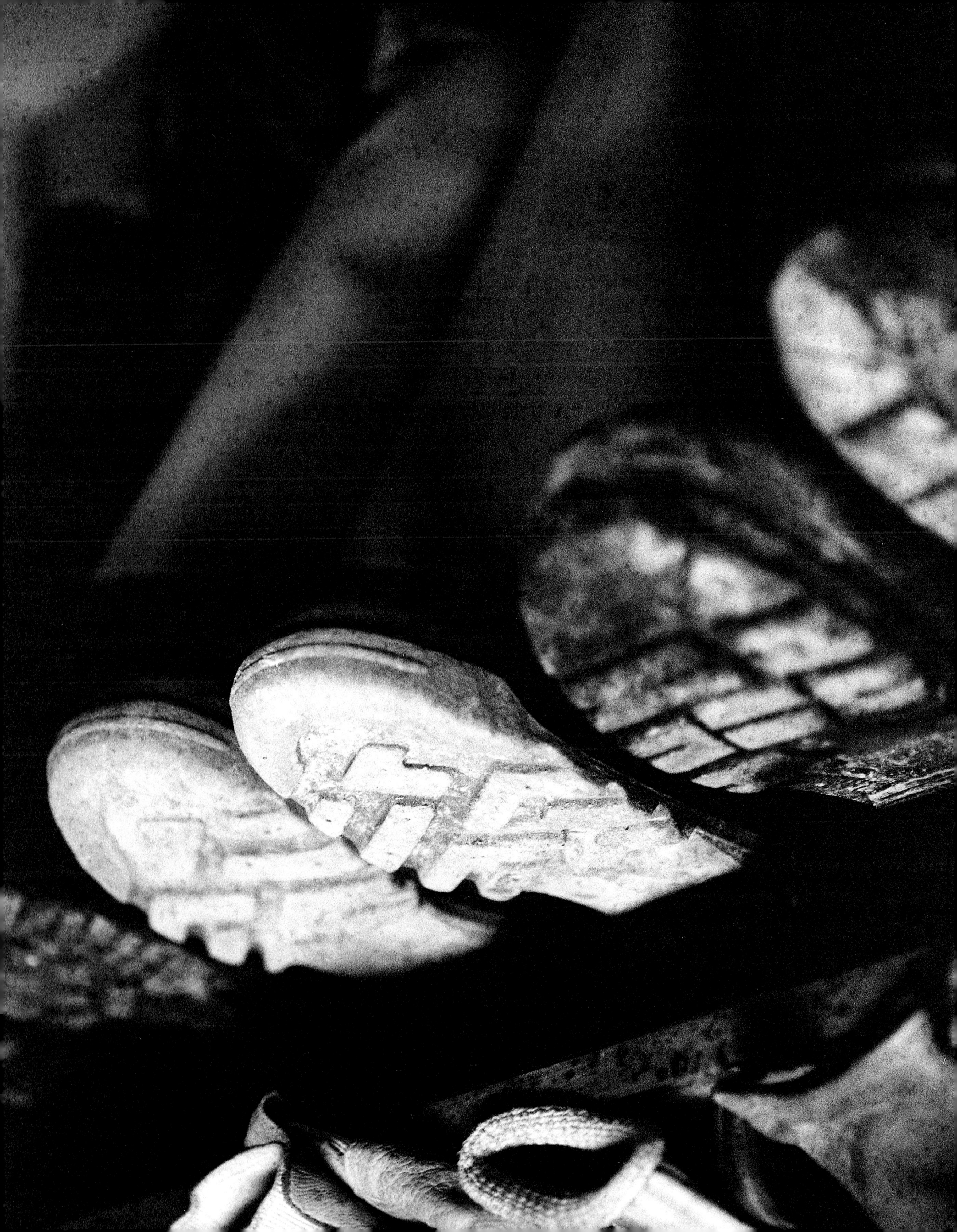

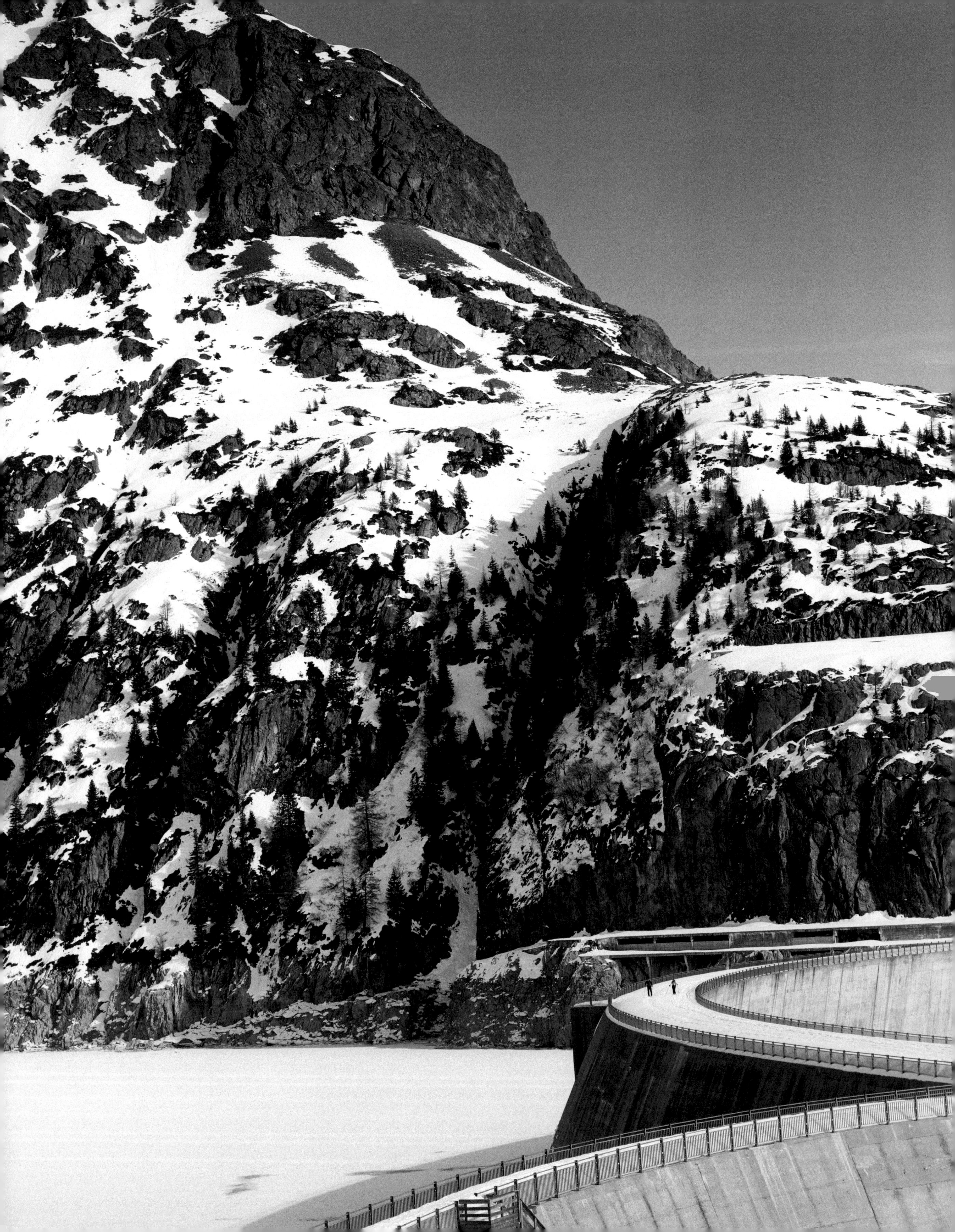

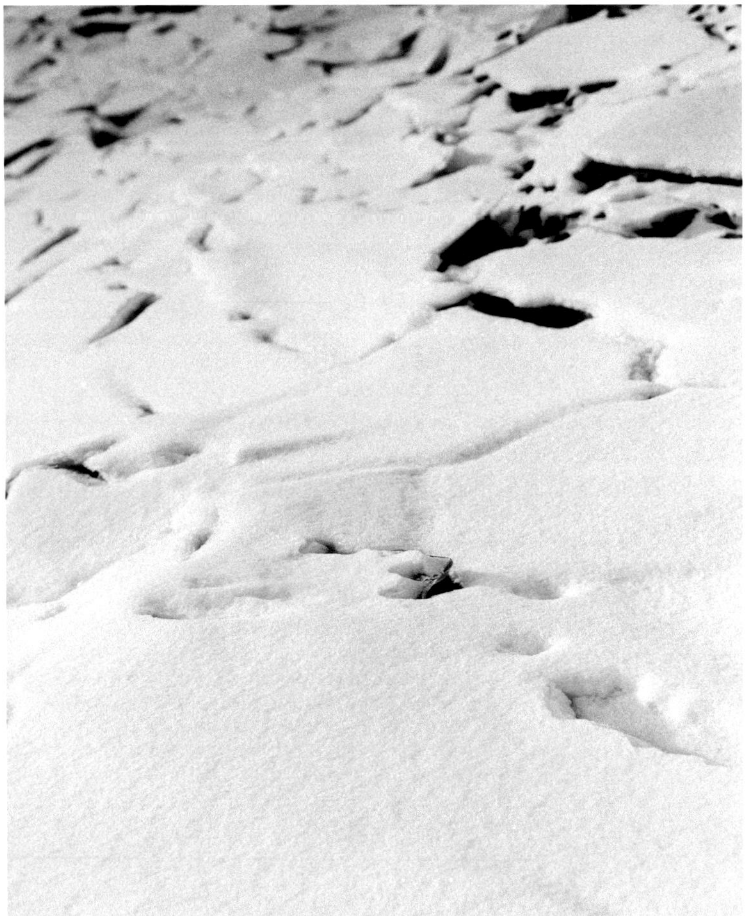

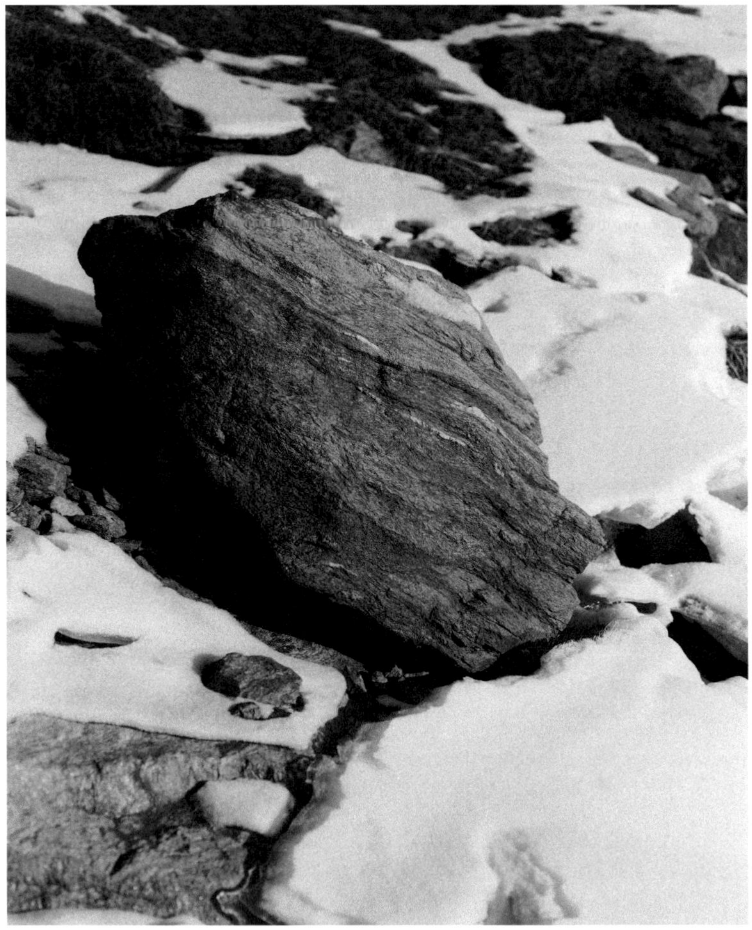

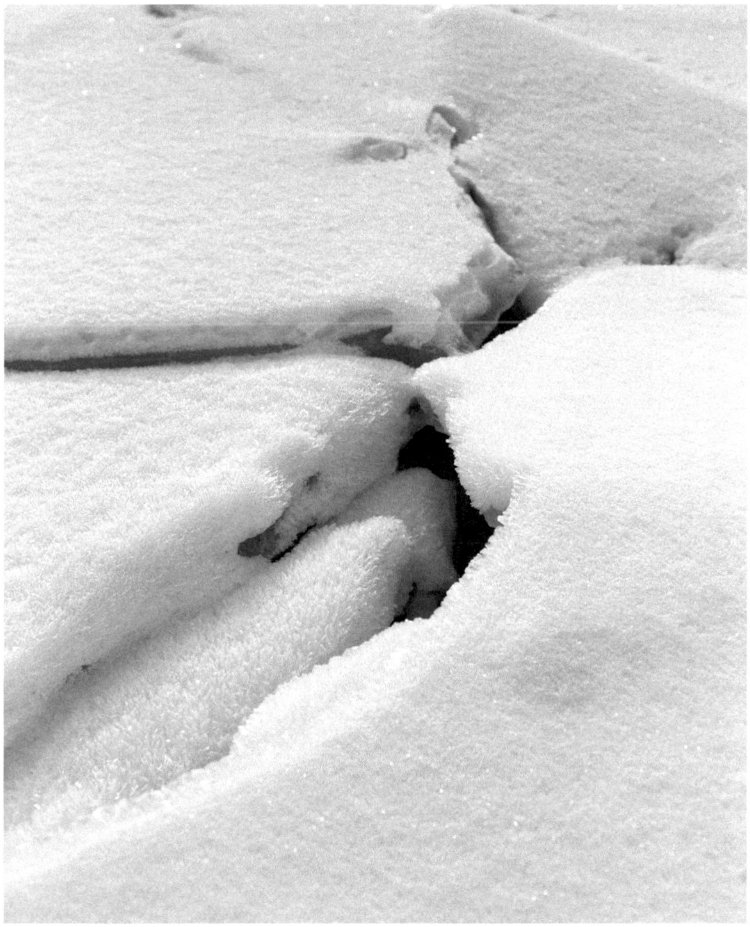
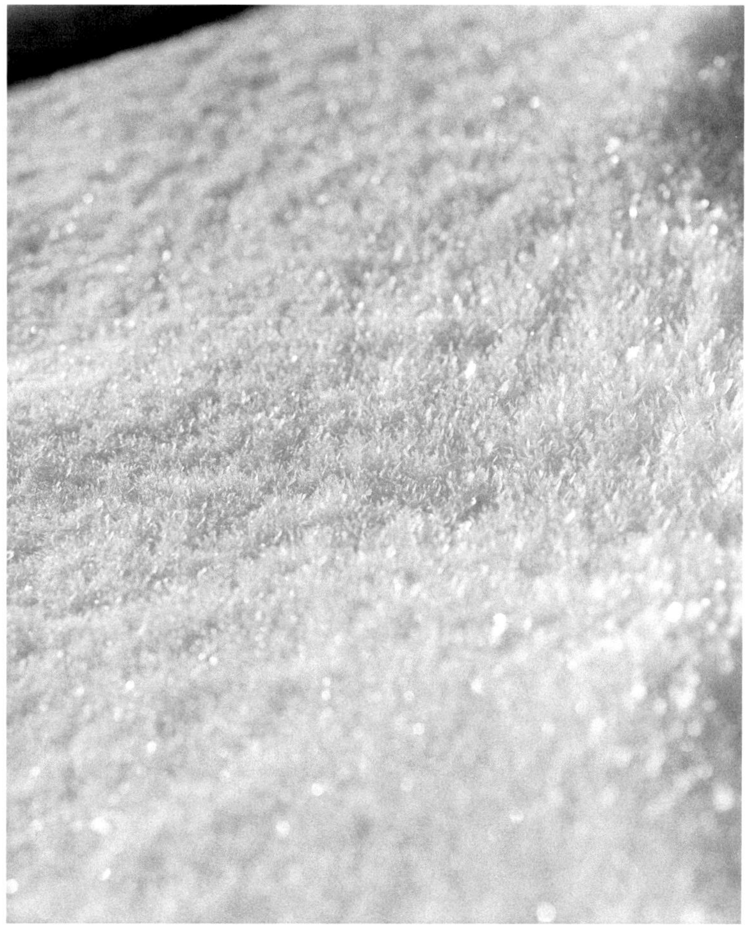

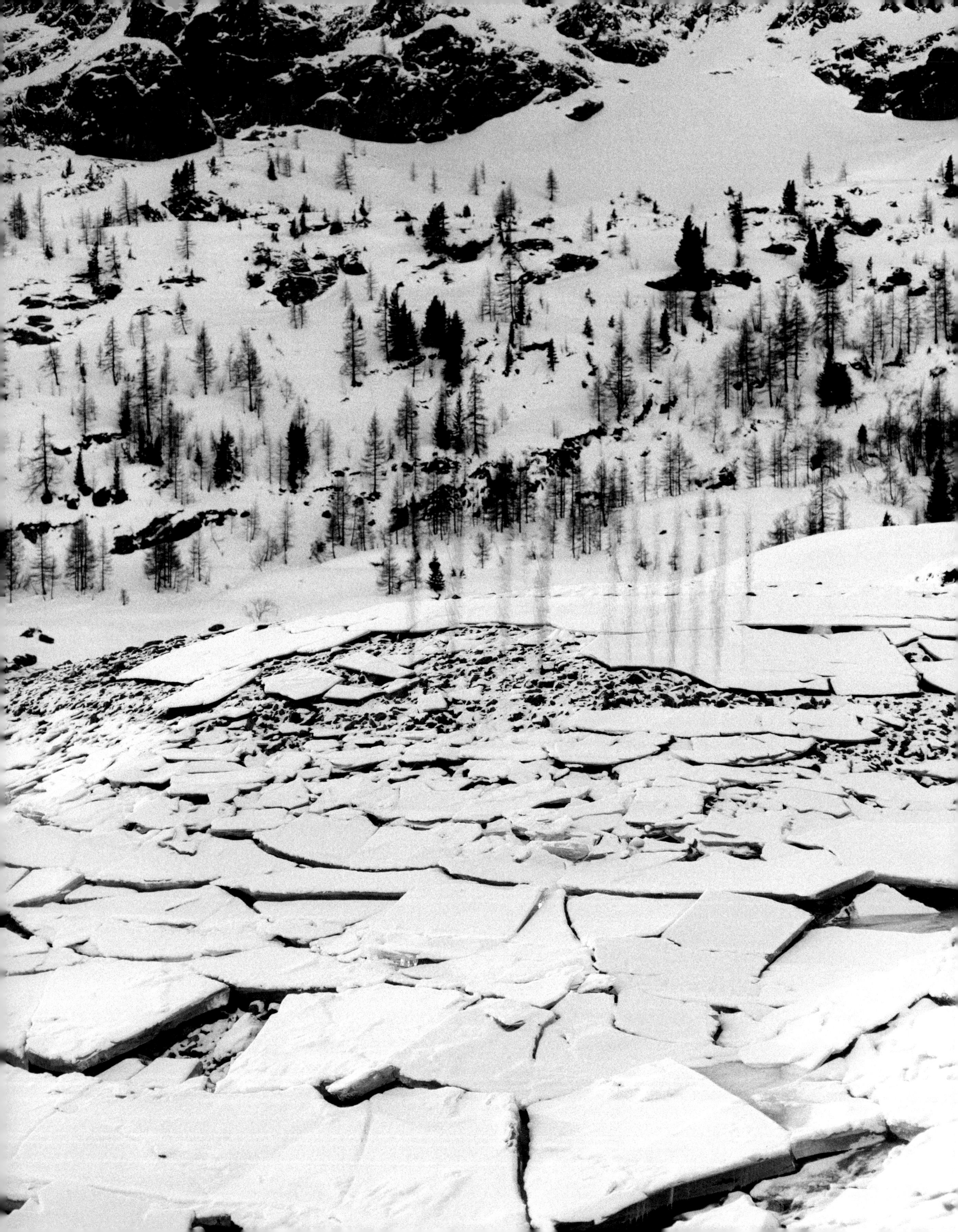

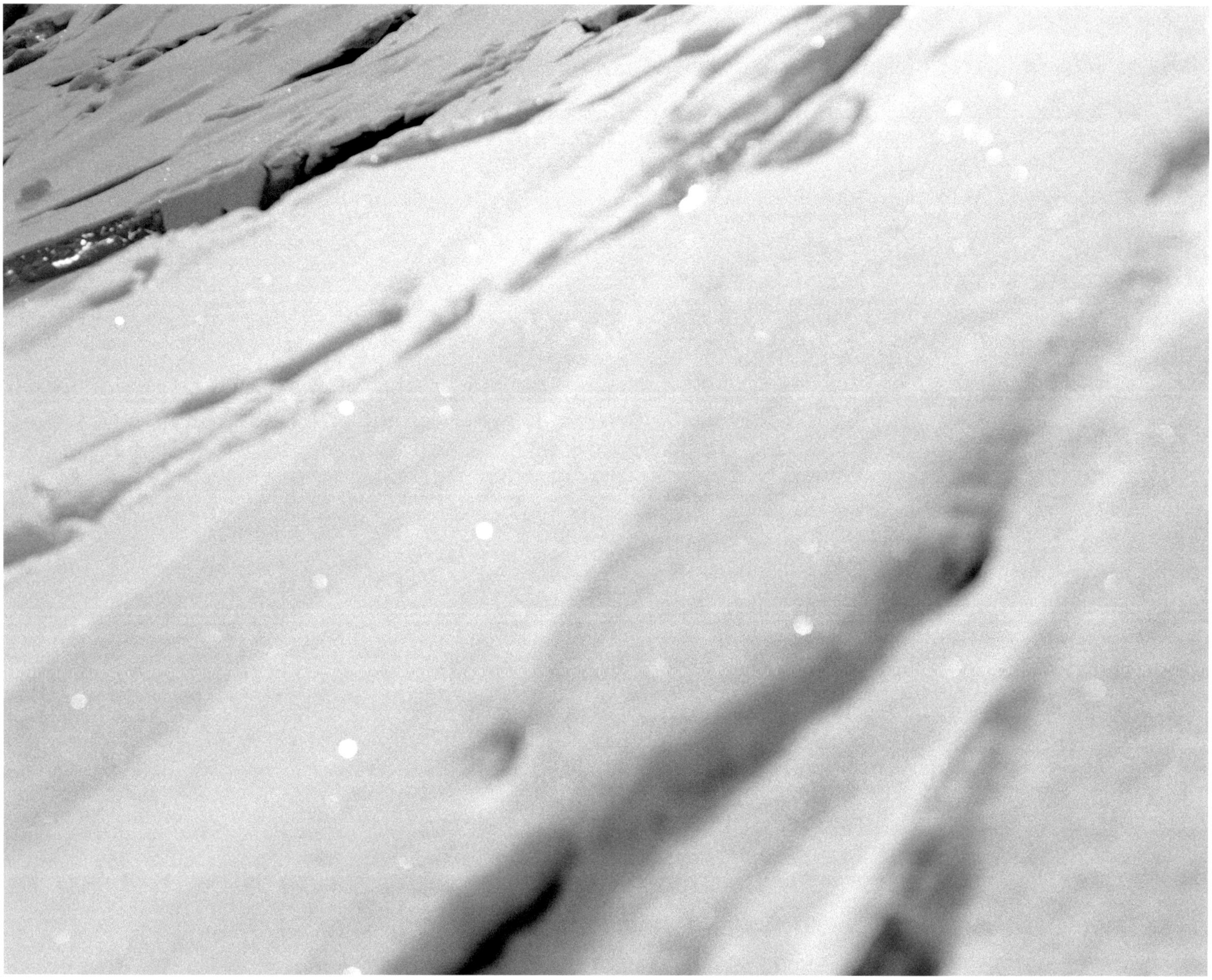

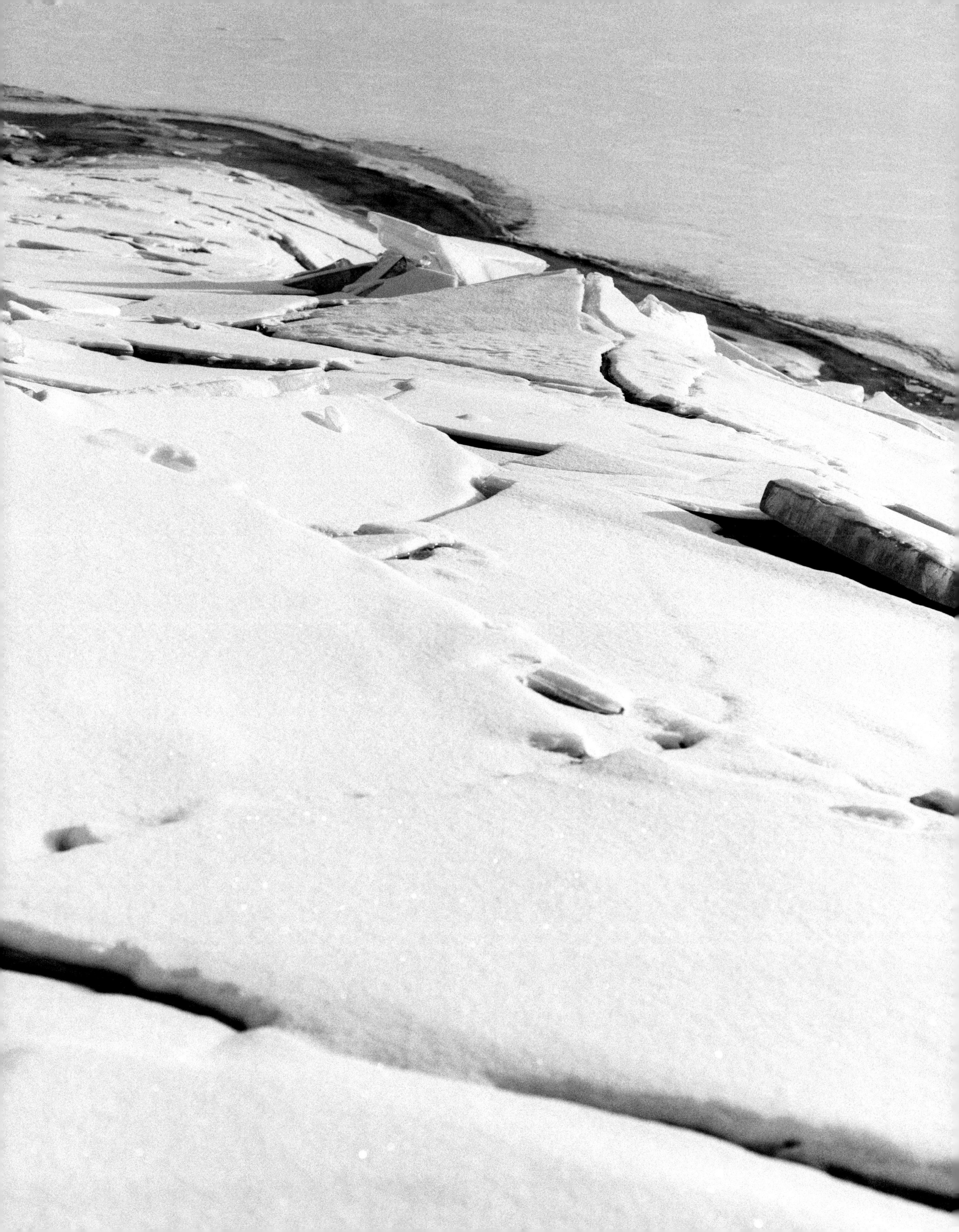

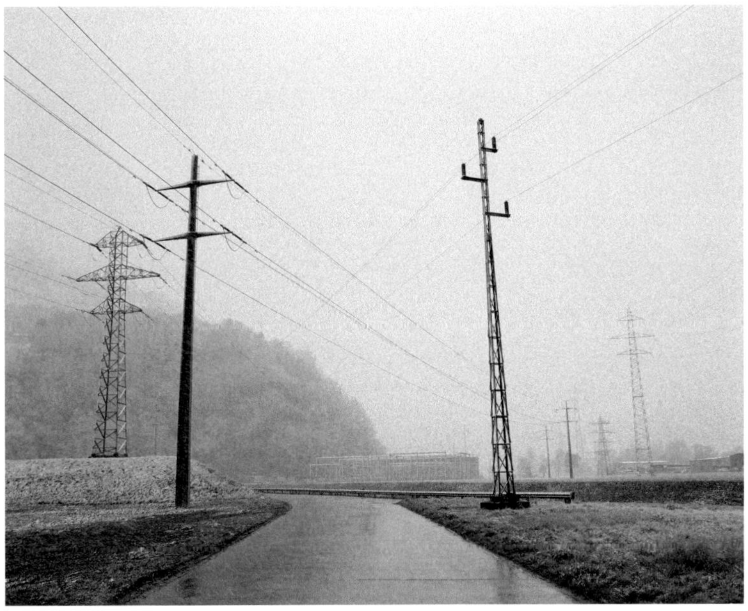

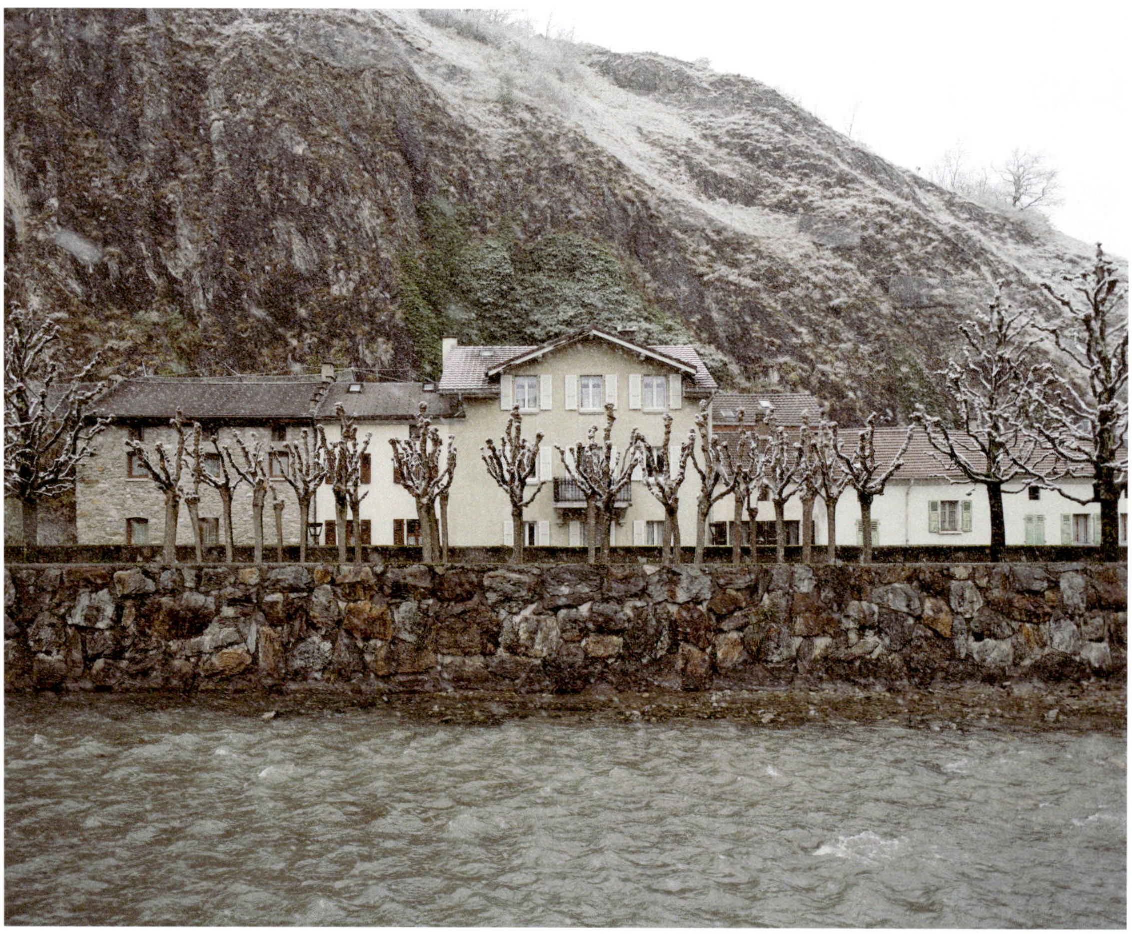

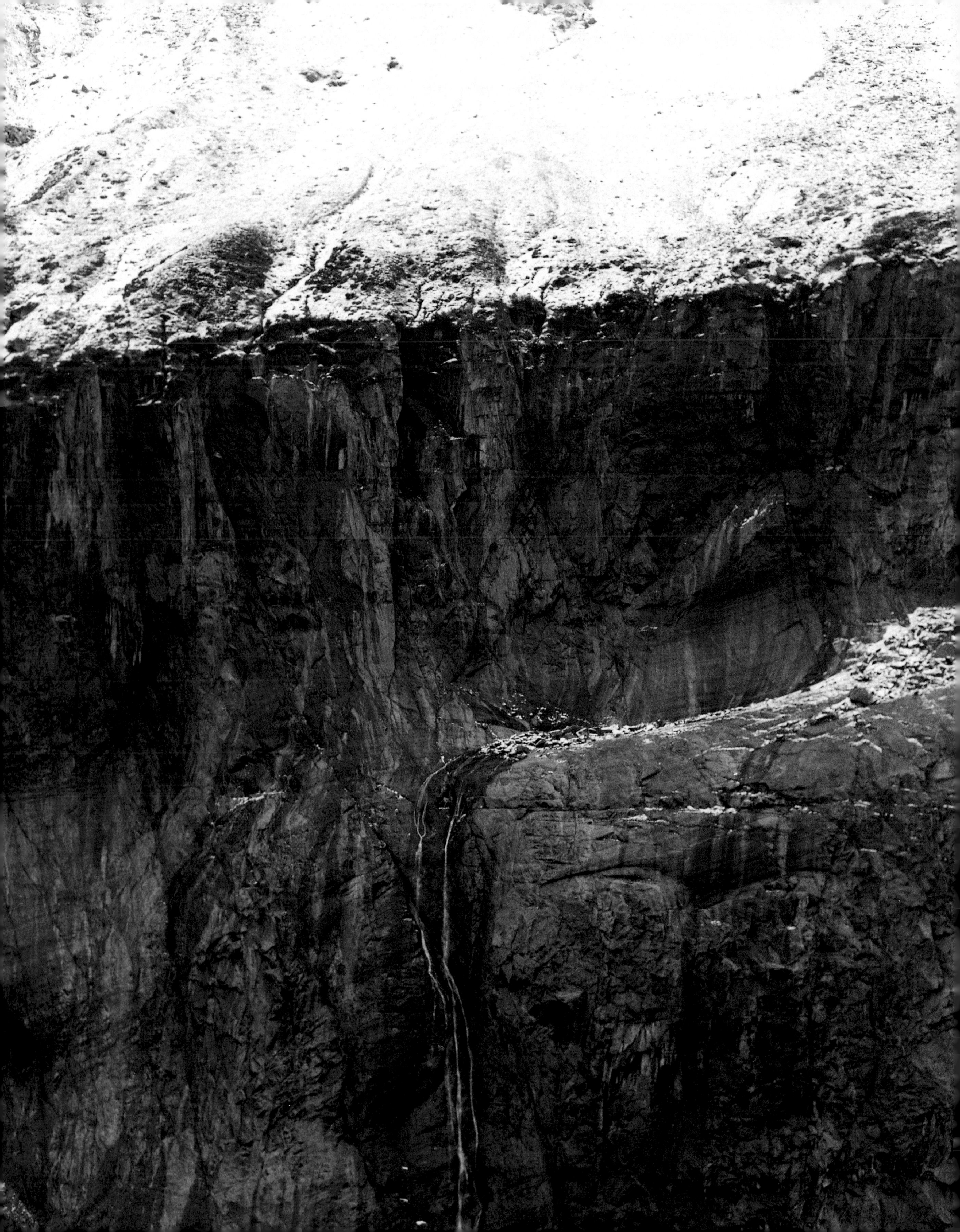

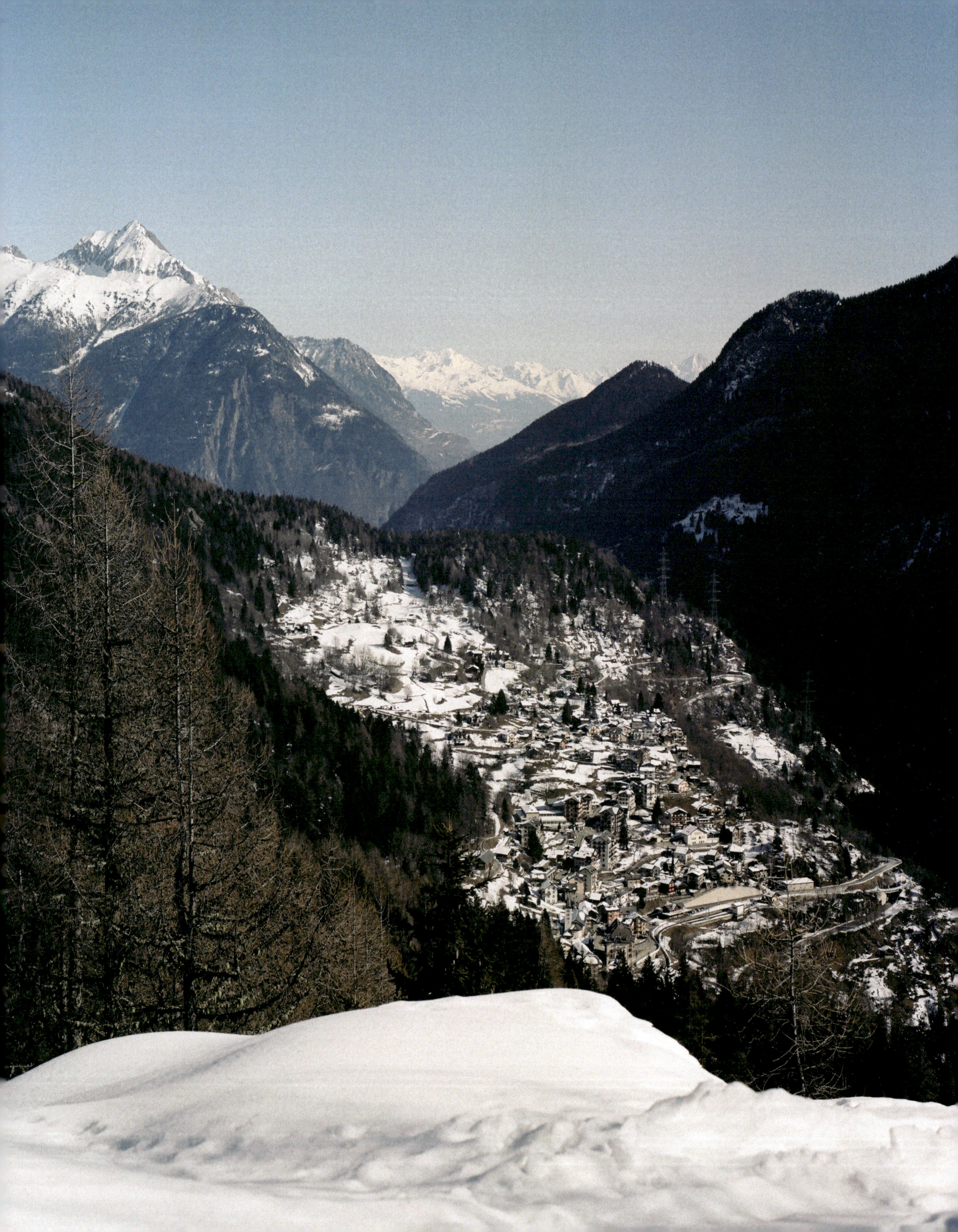

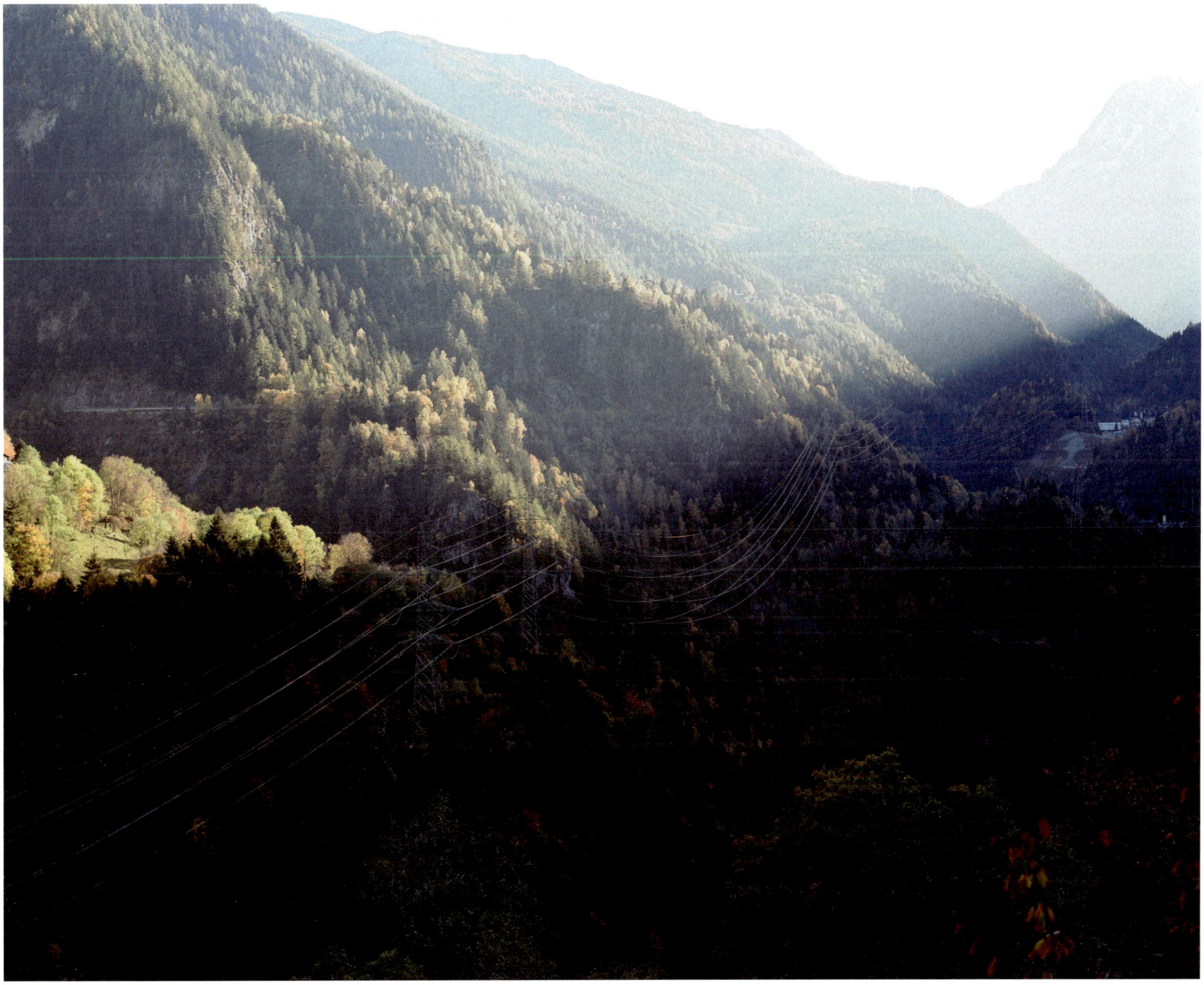

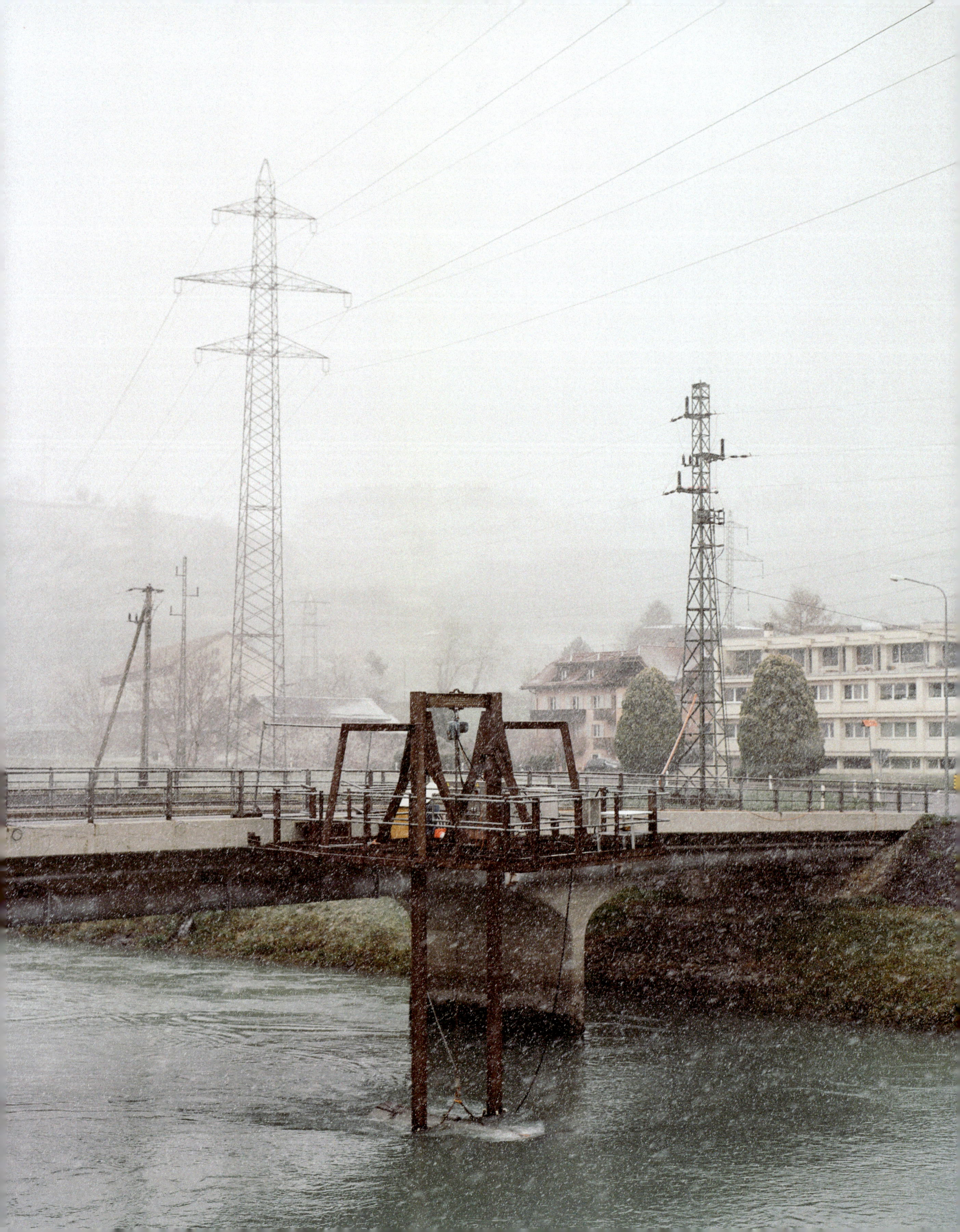

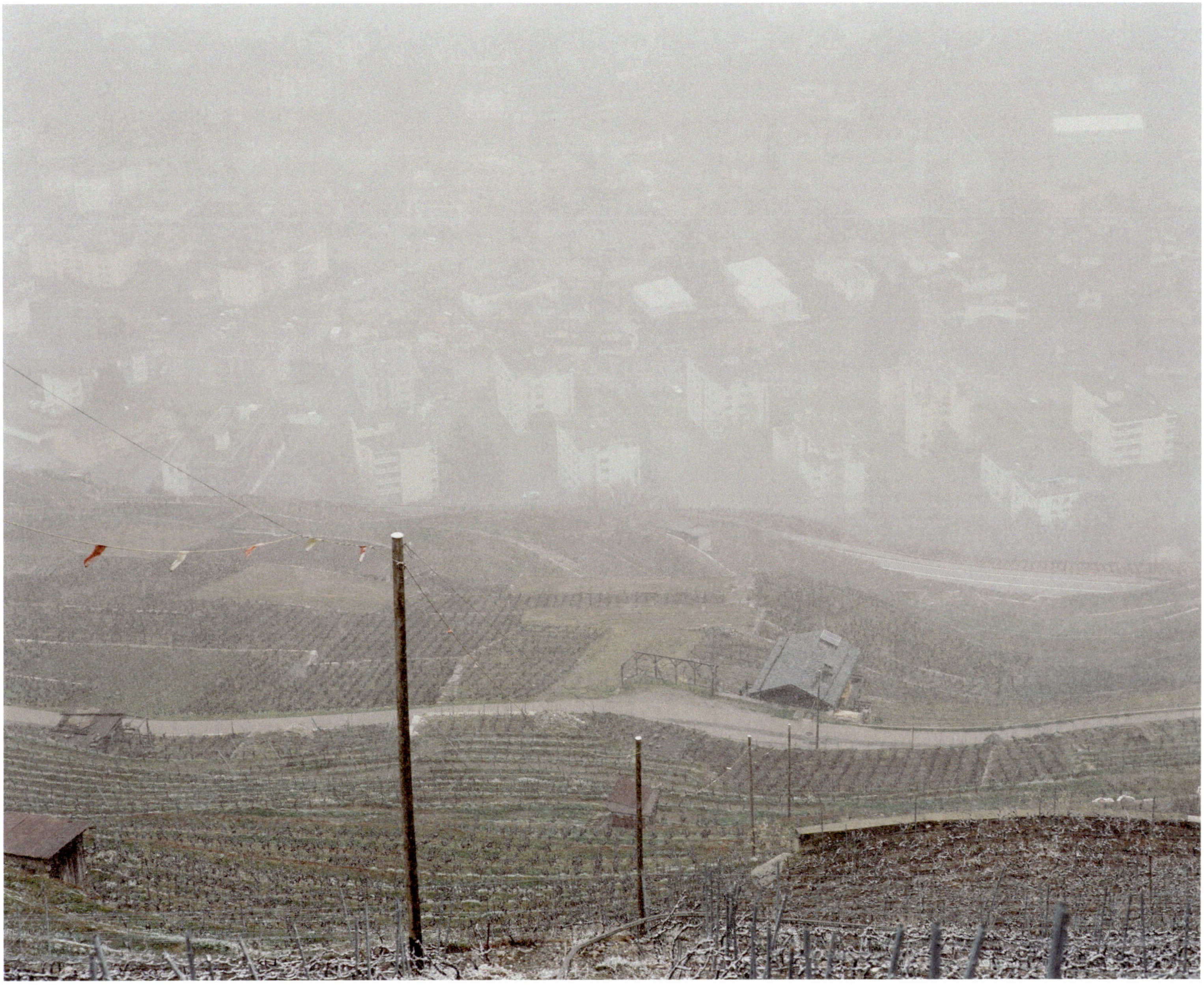

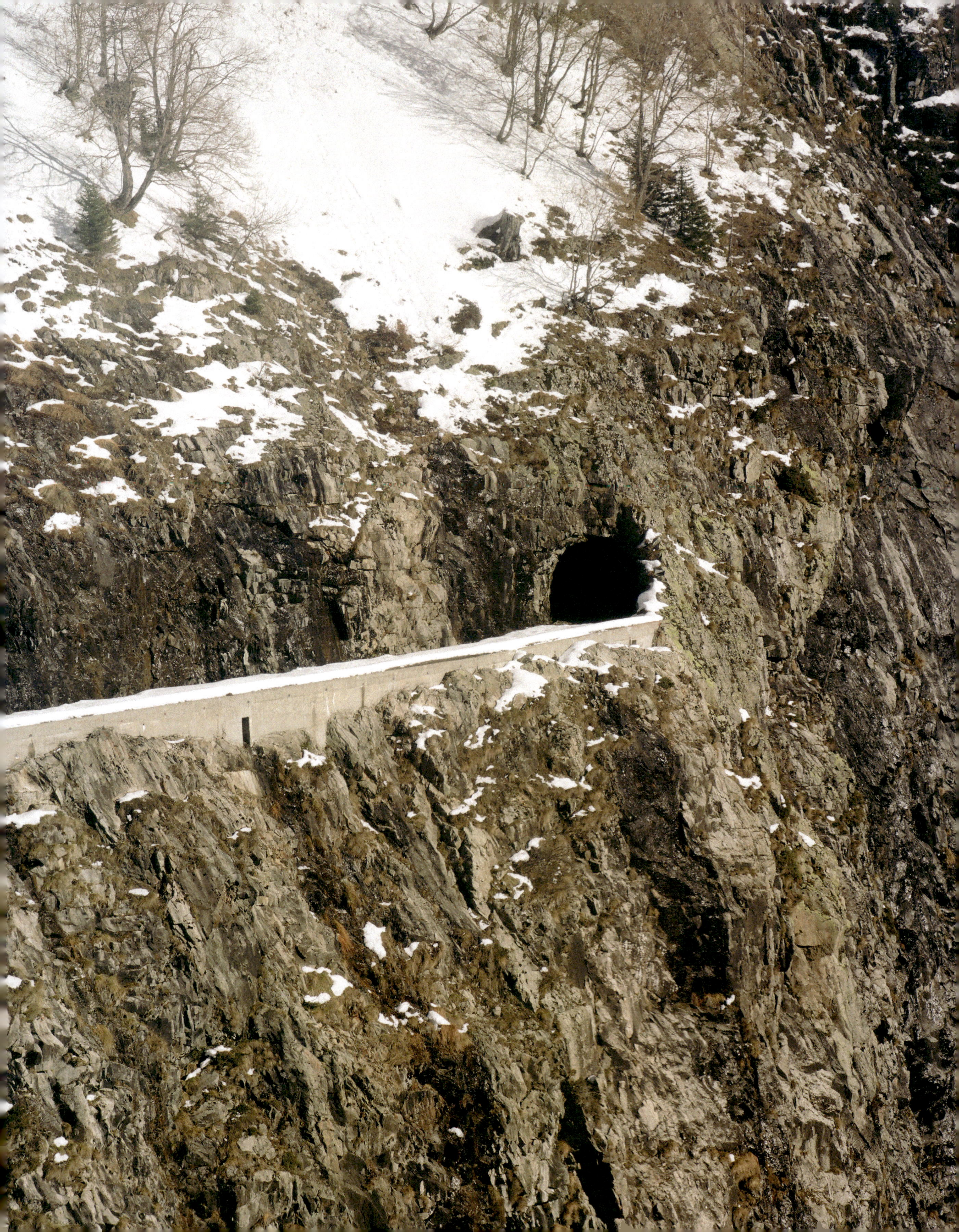

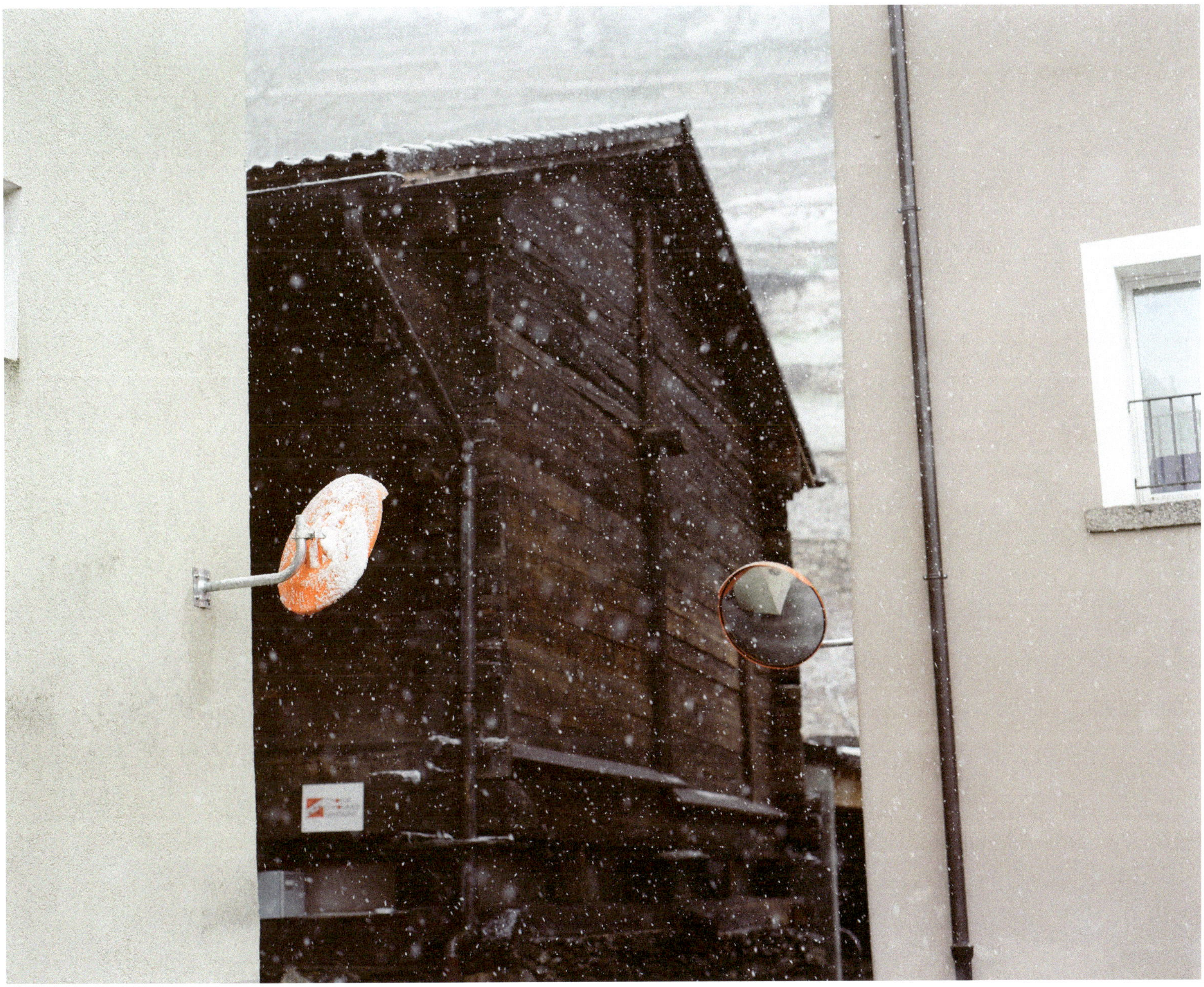

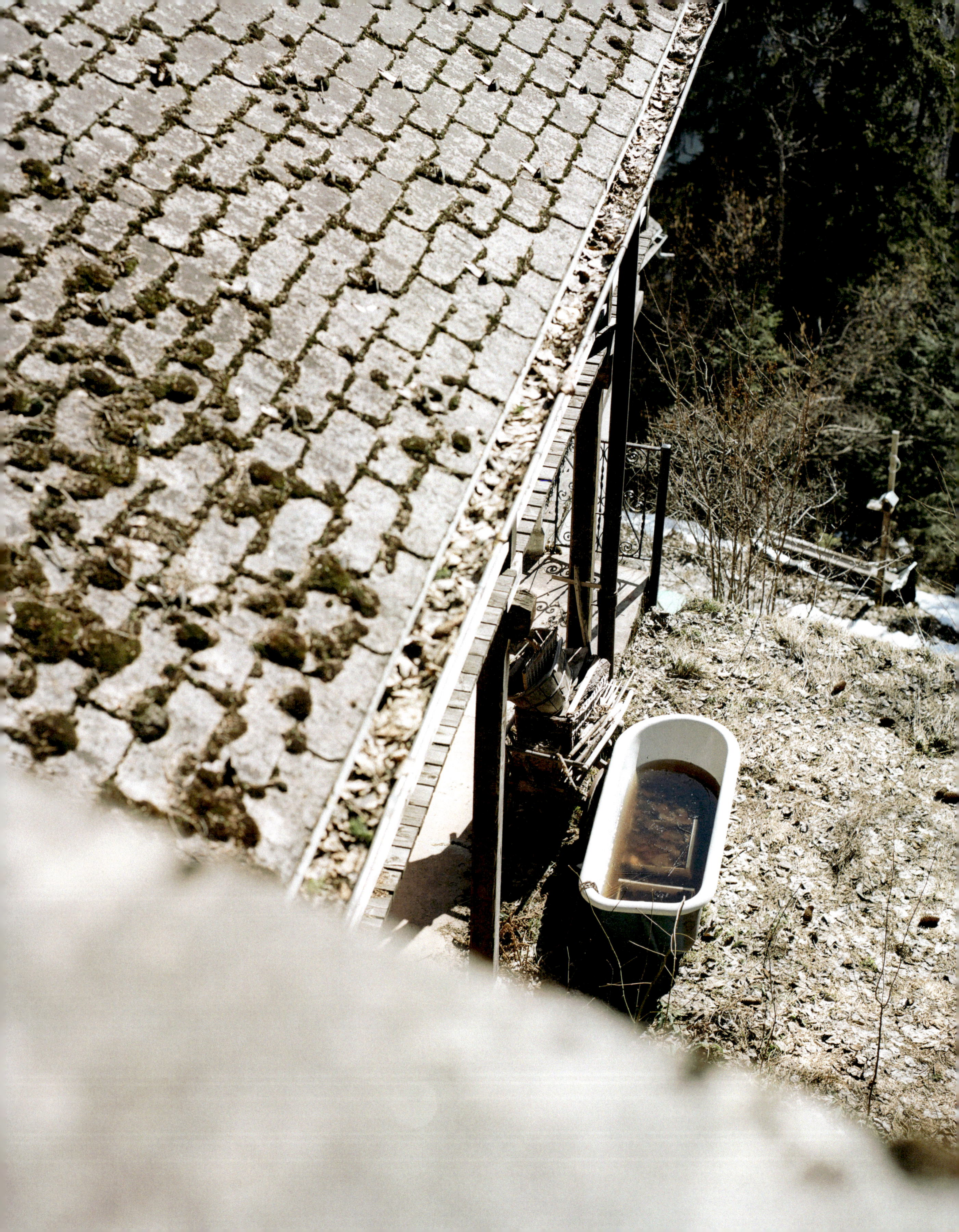

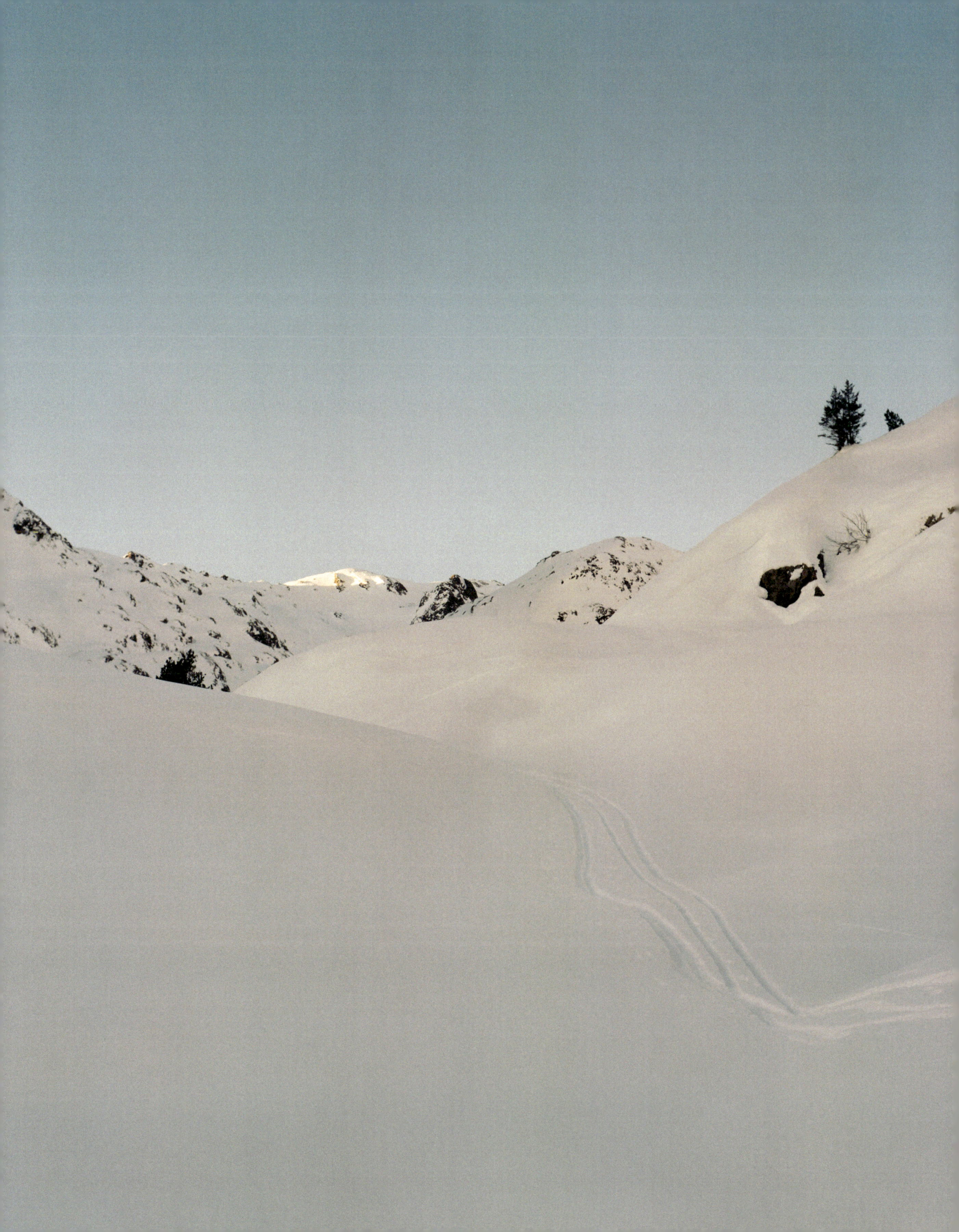

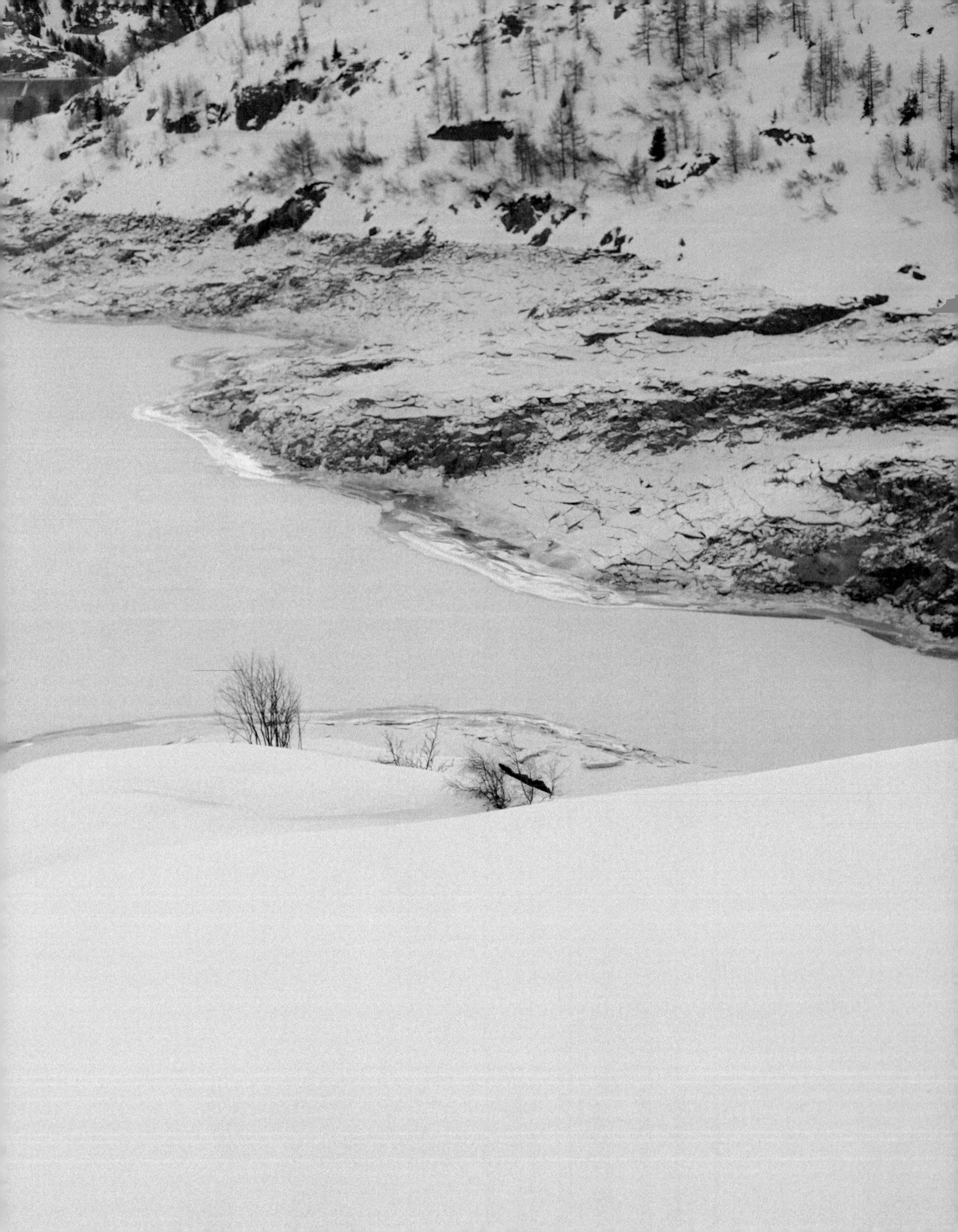

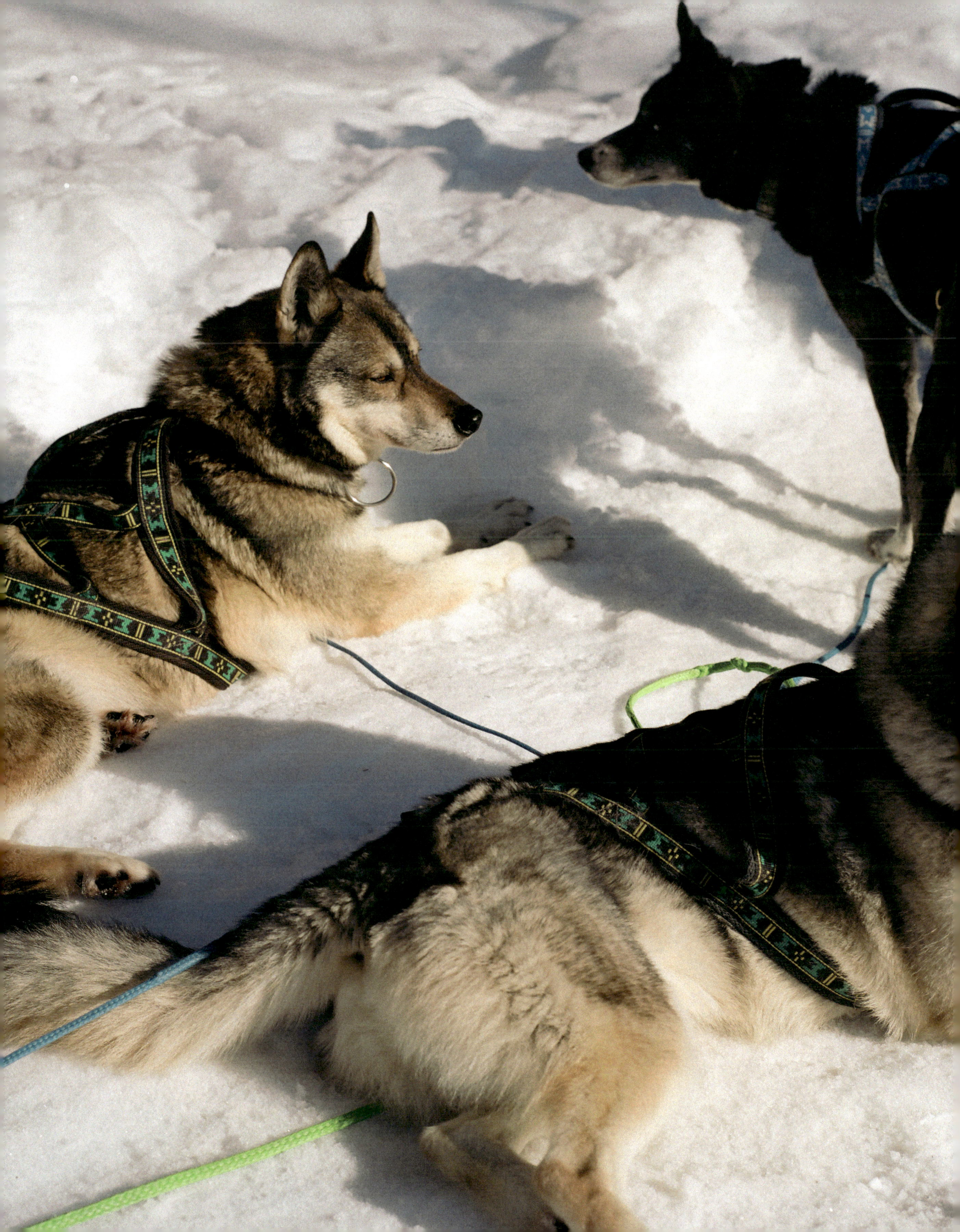

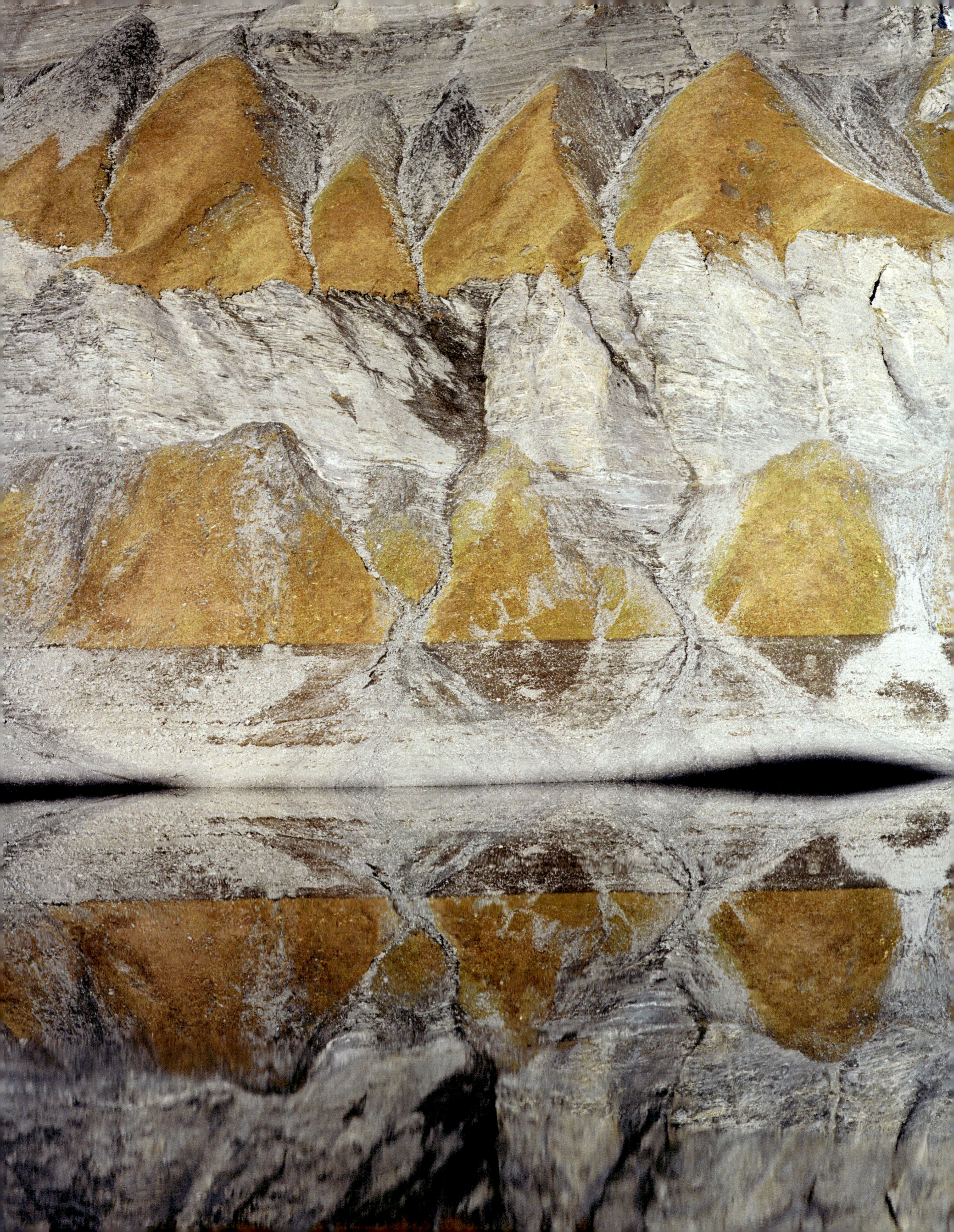

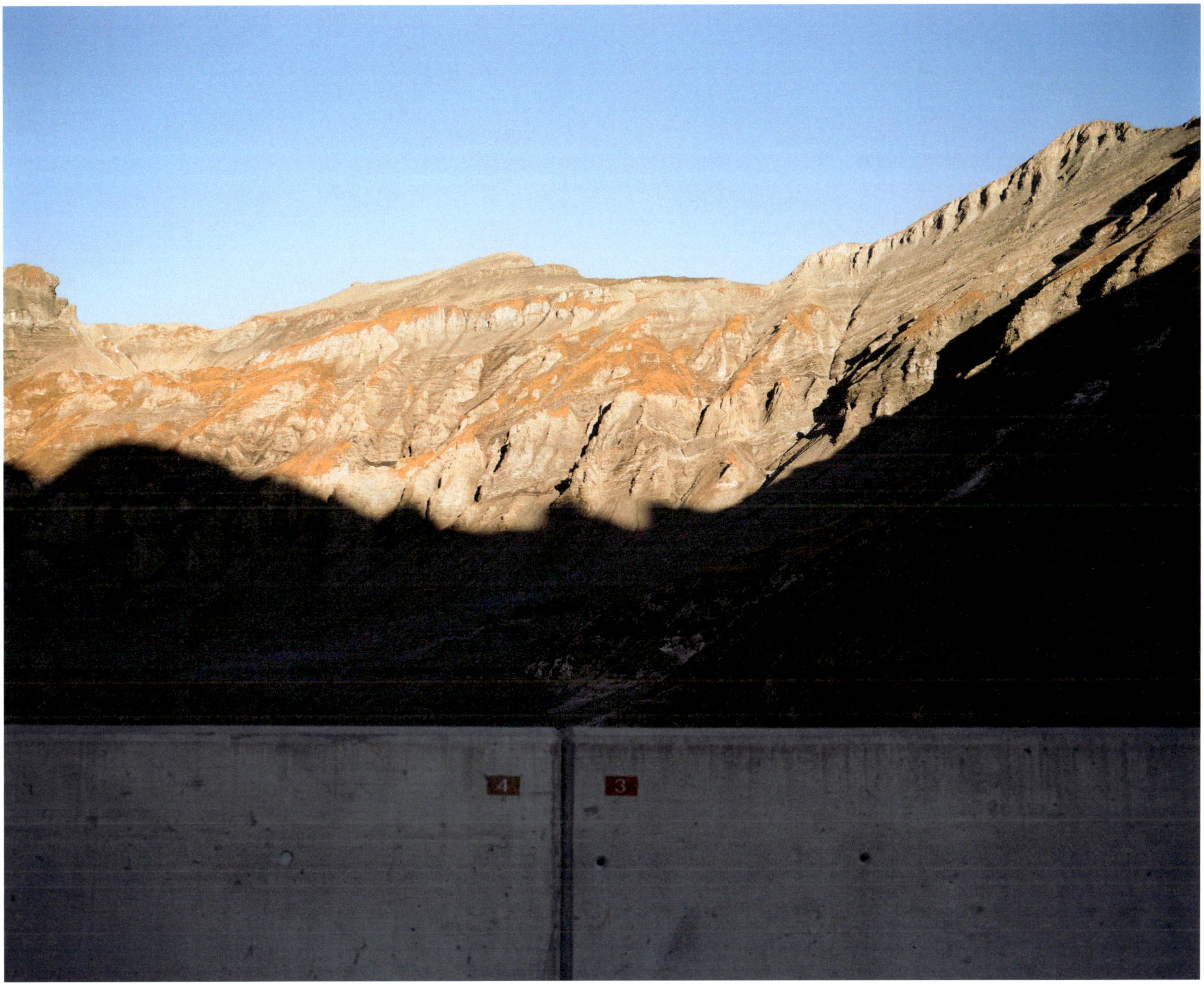
4
3

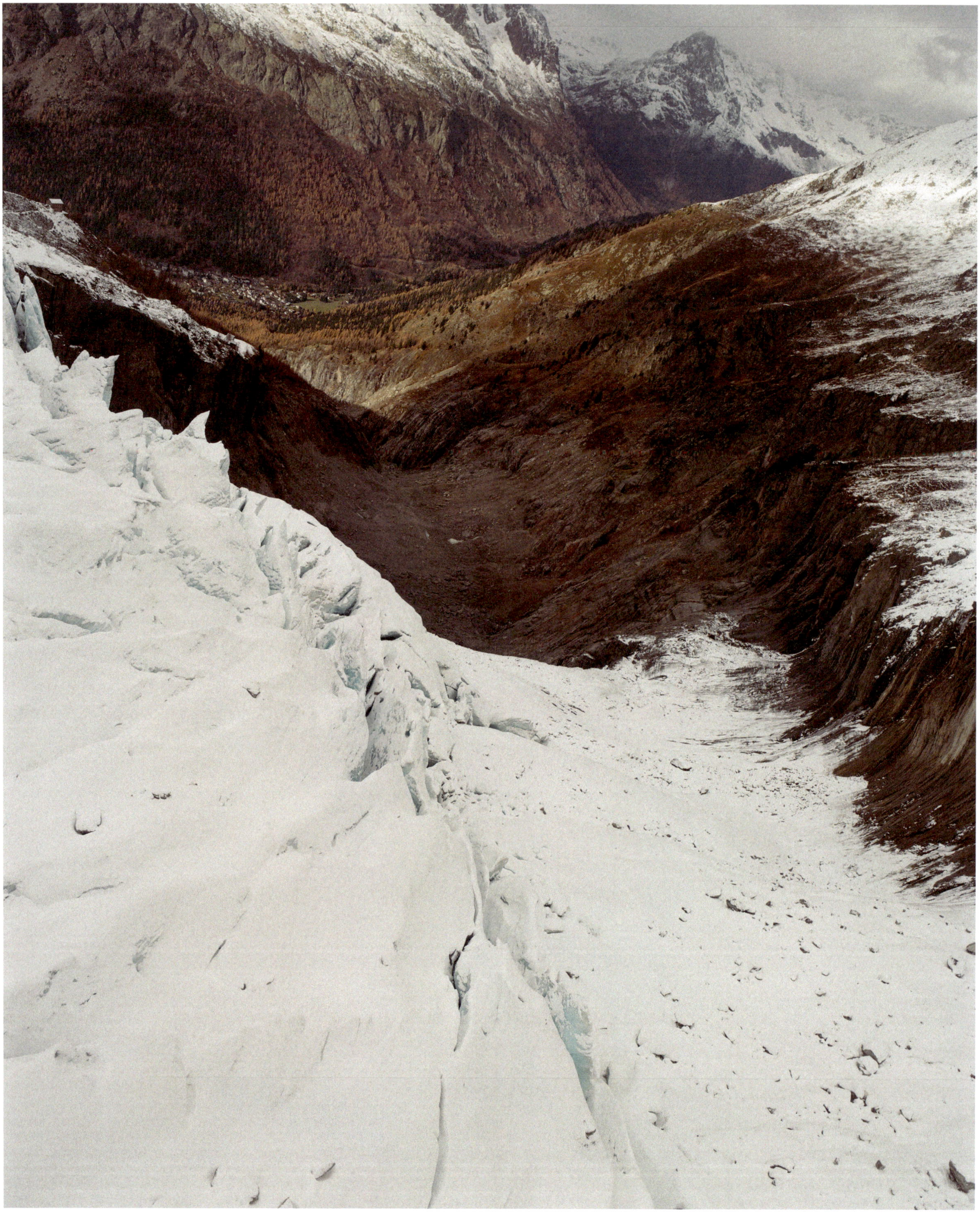

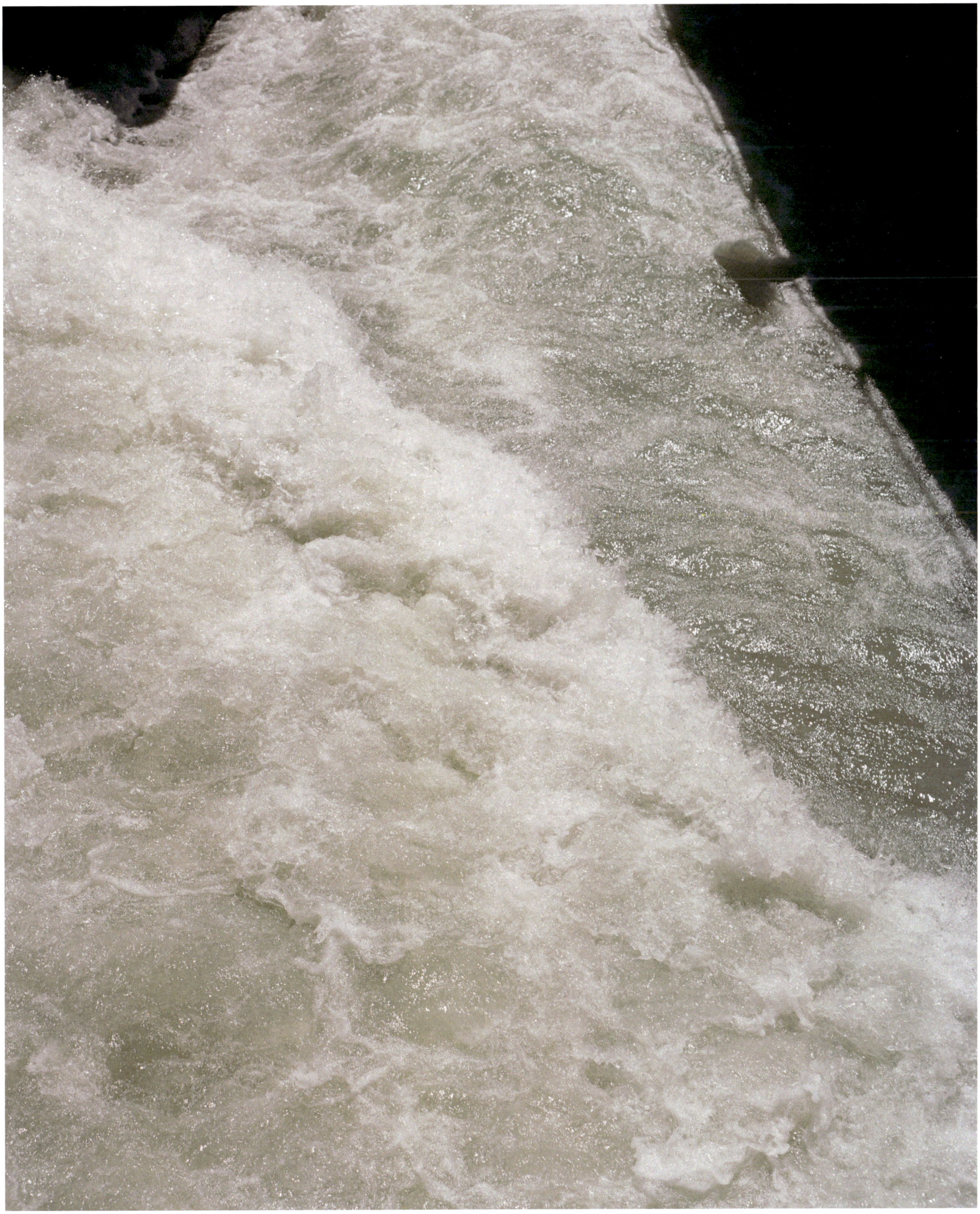

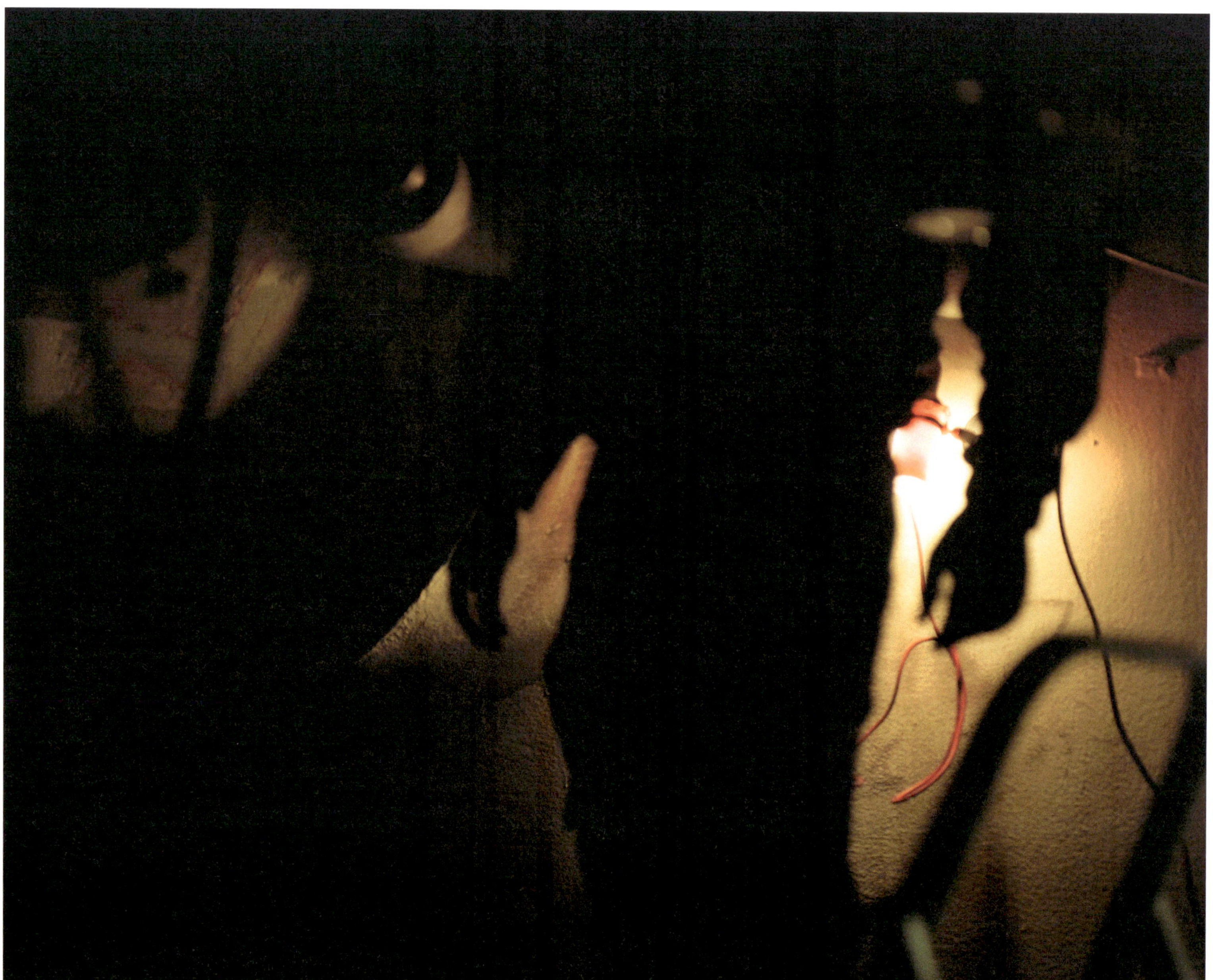

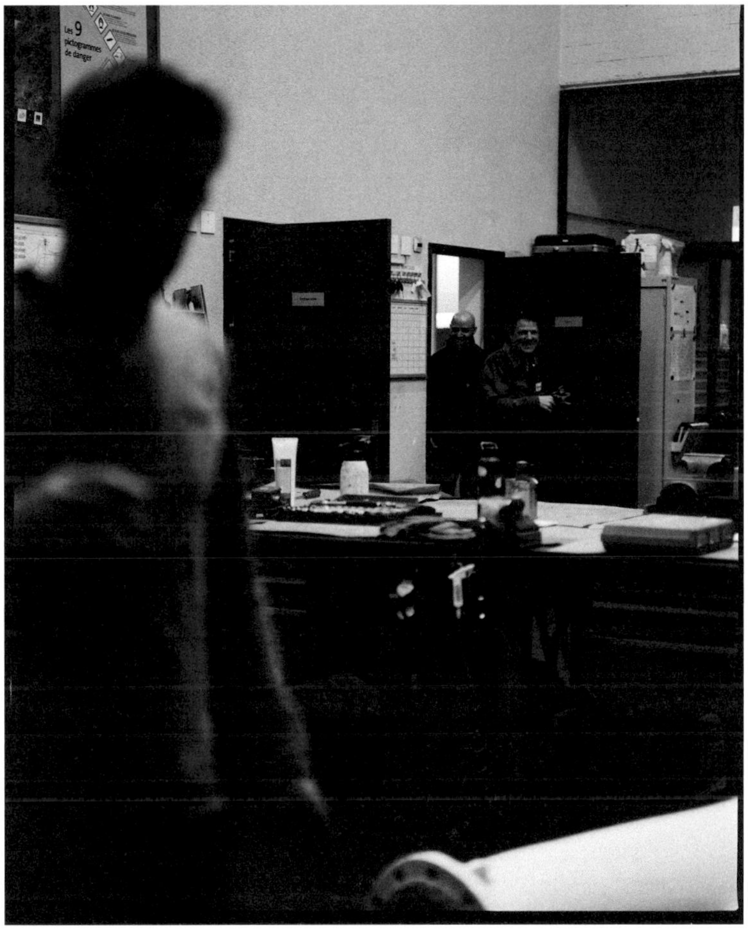
Les 9
pictogrammes
de danger

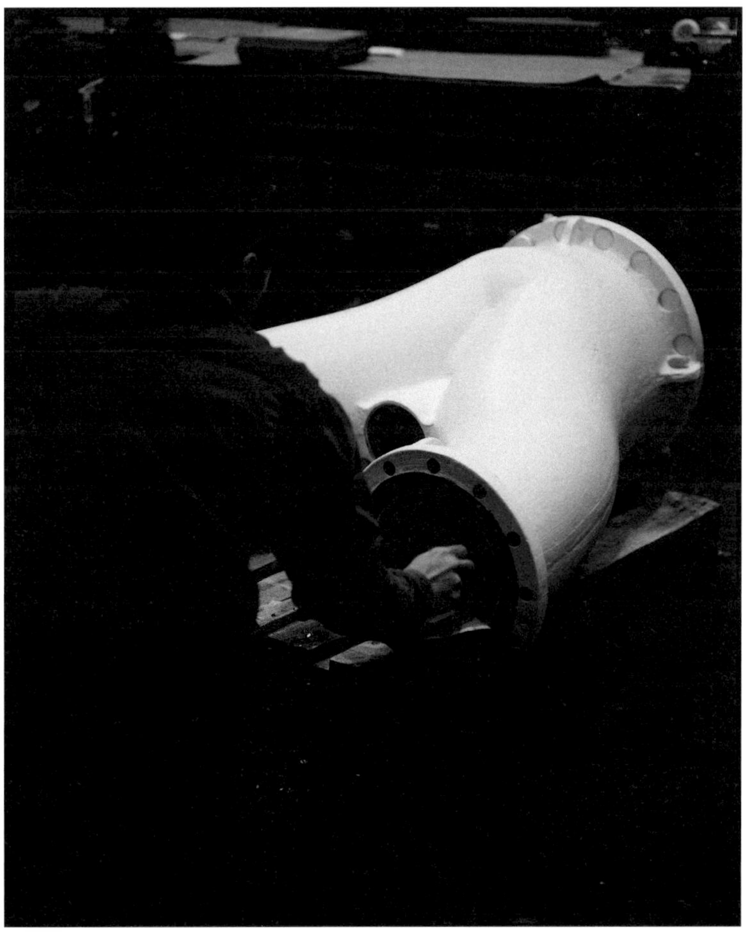

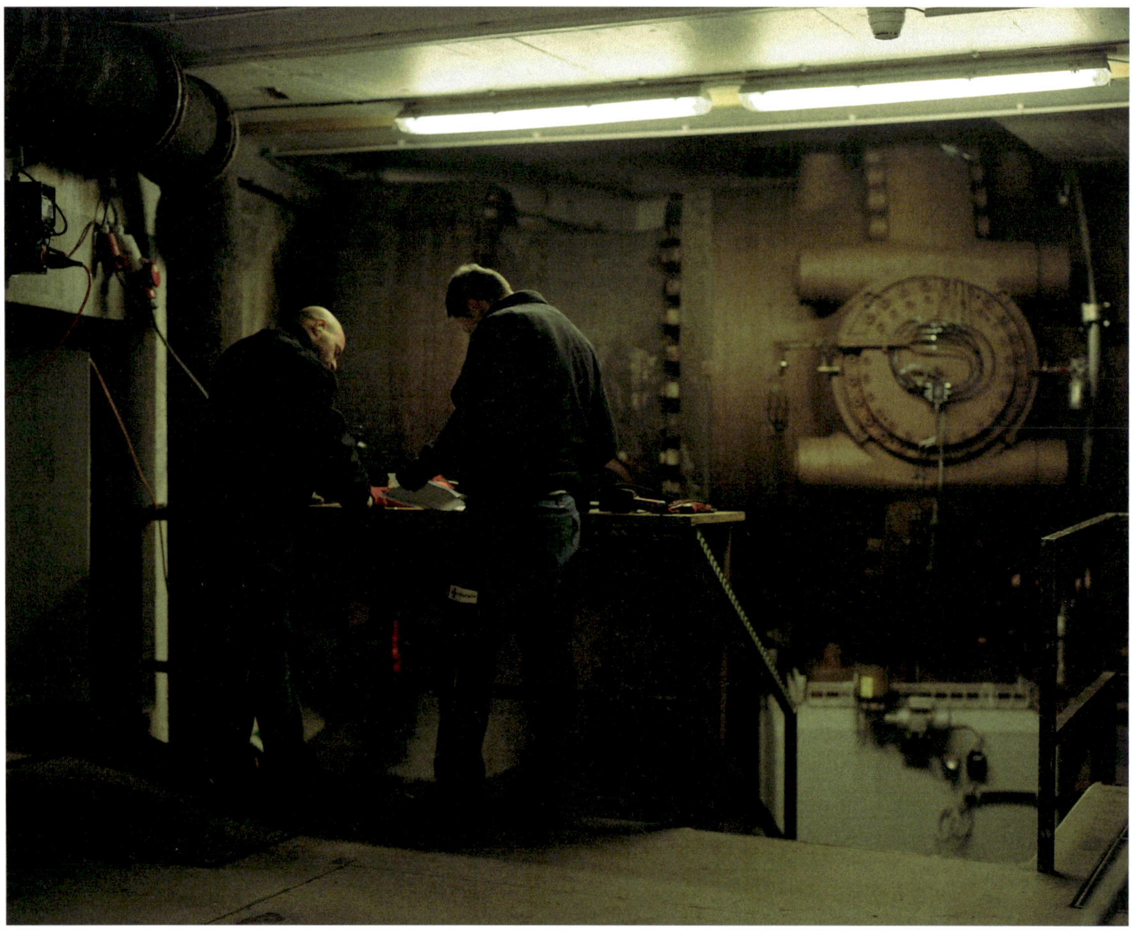

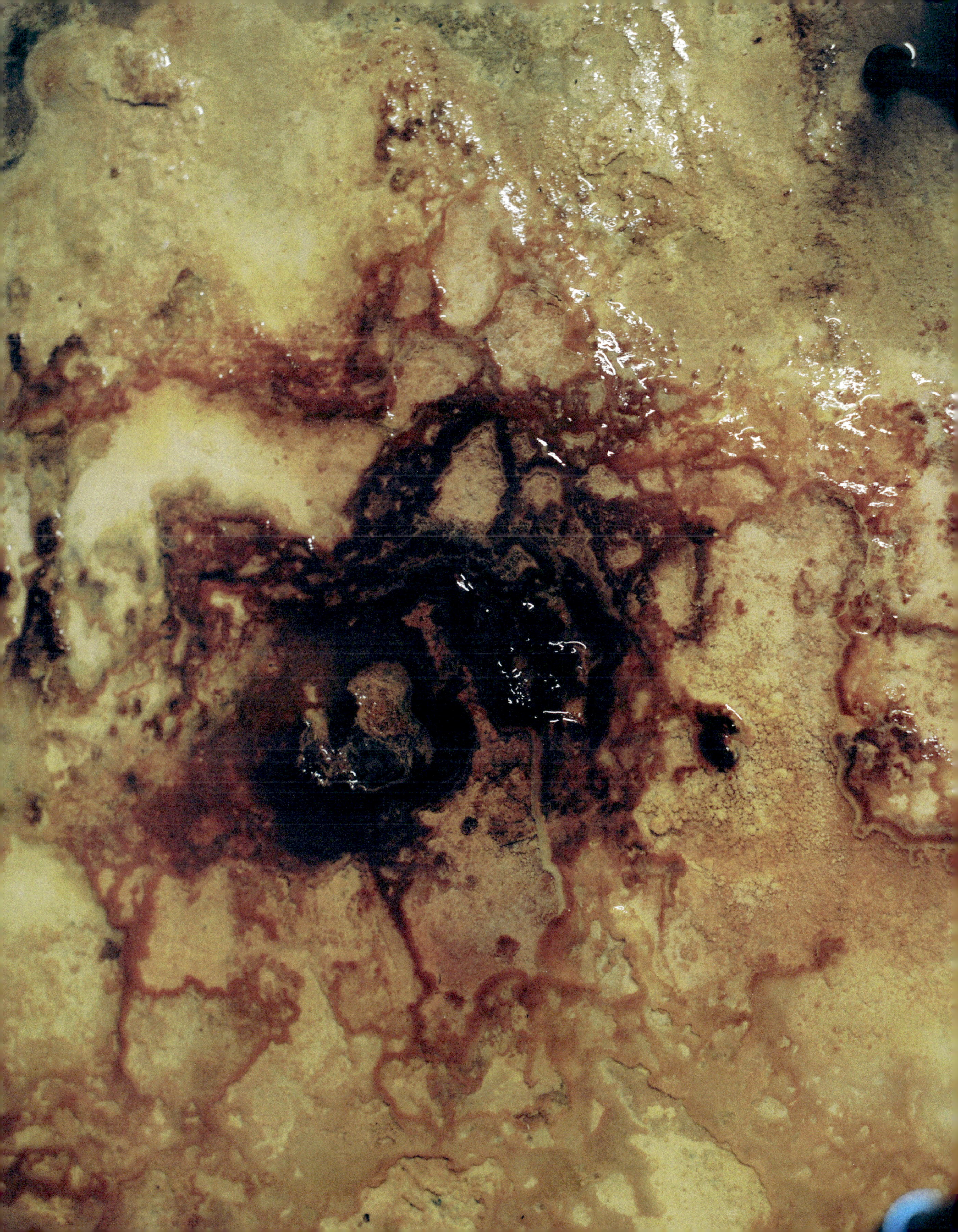

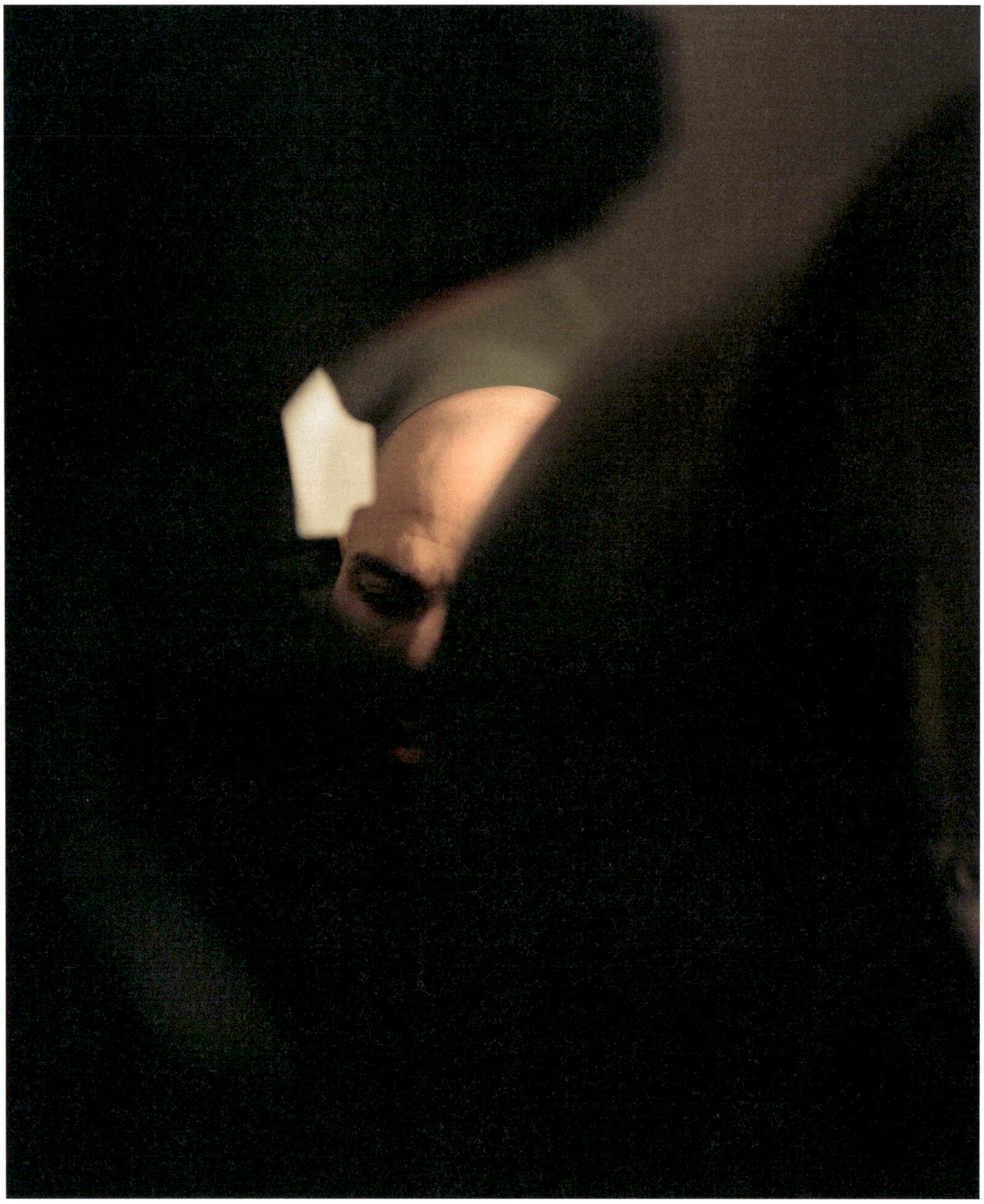

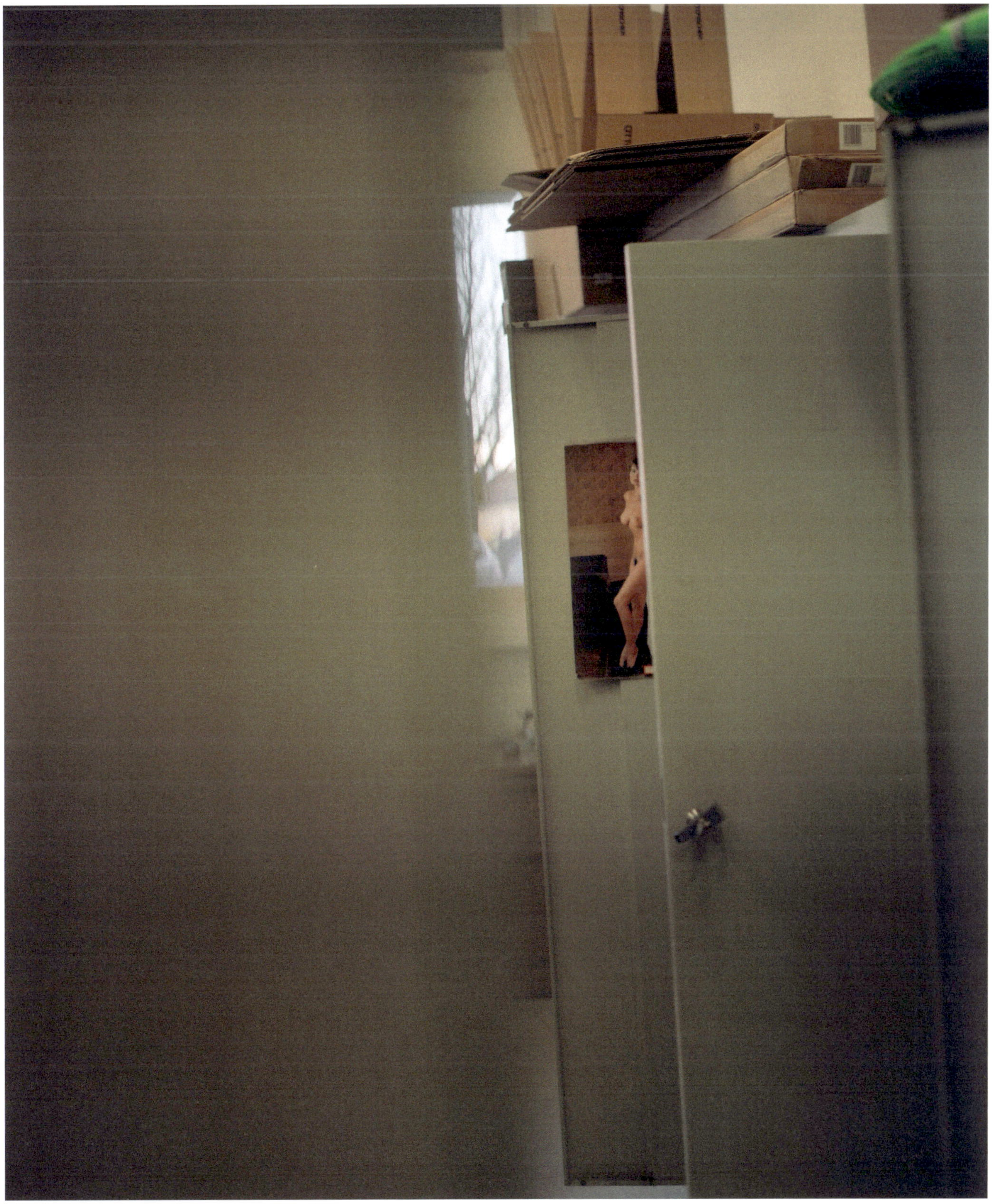

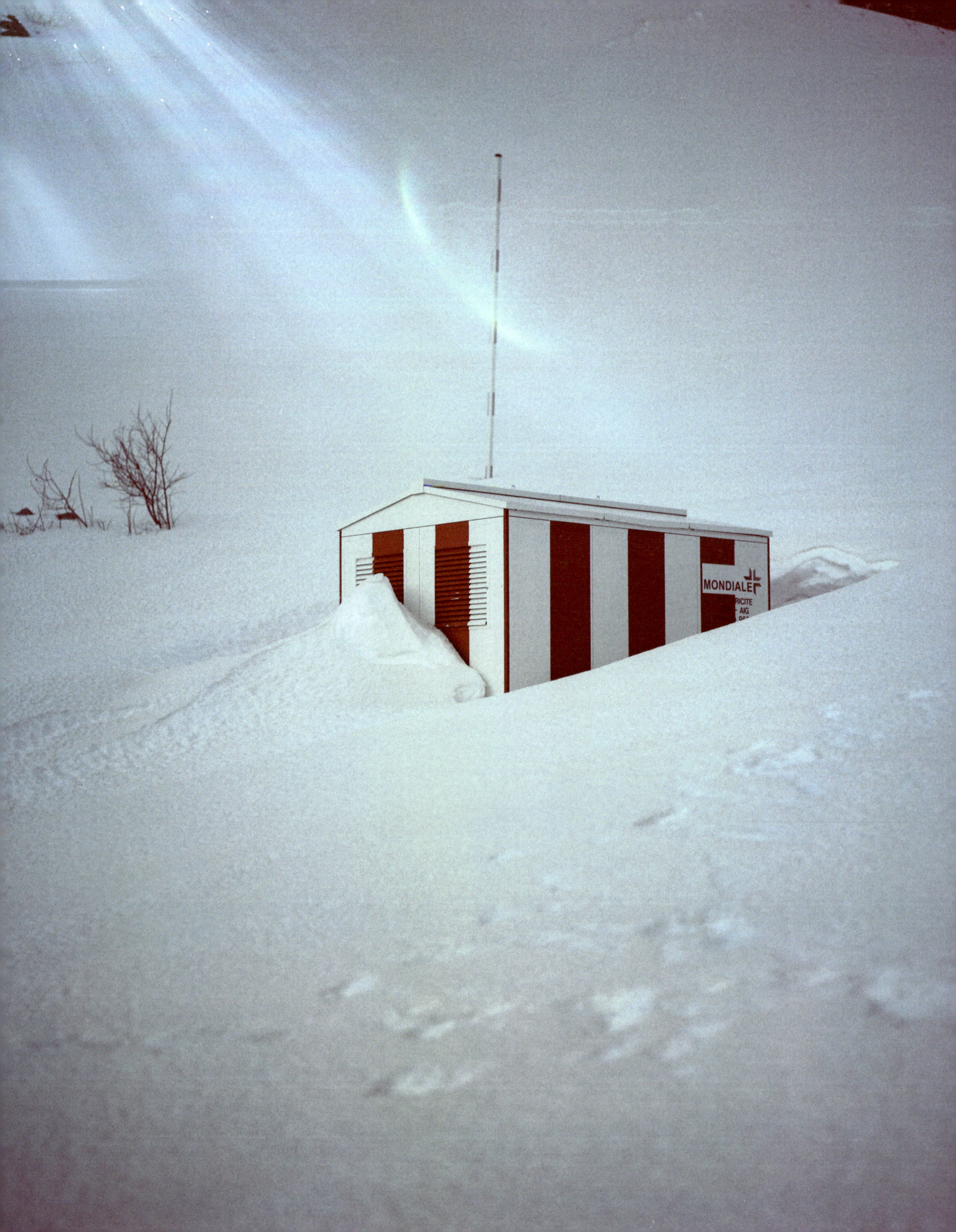
MONDIALE

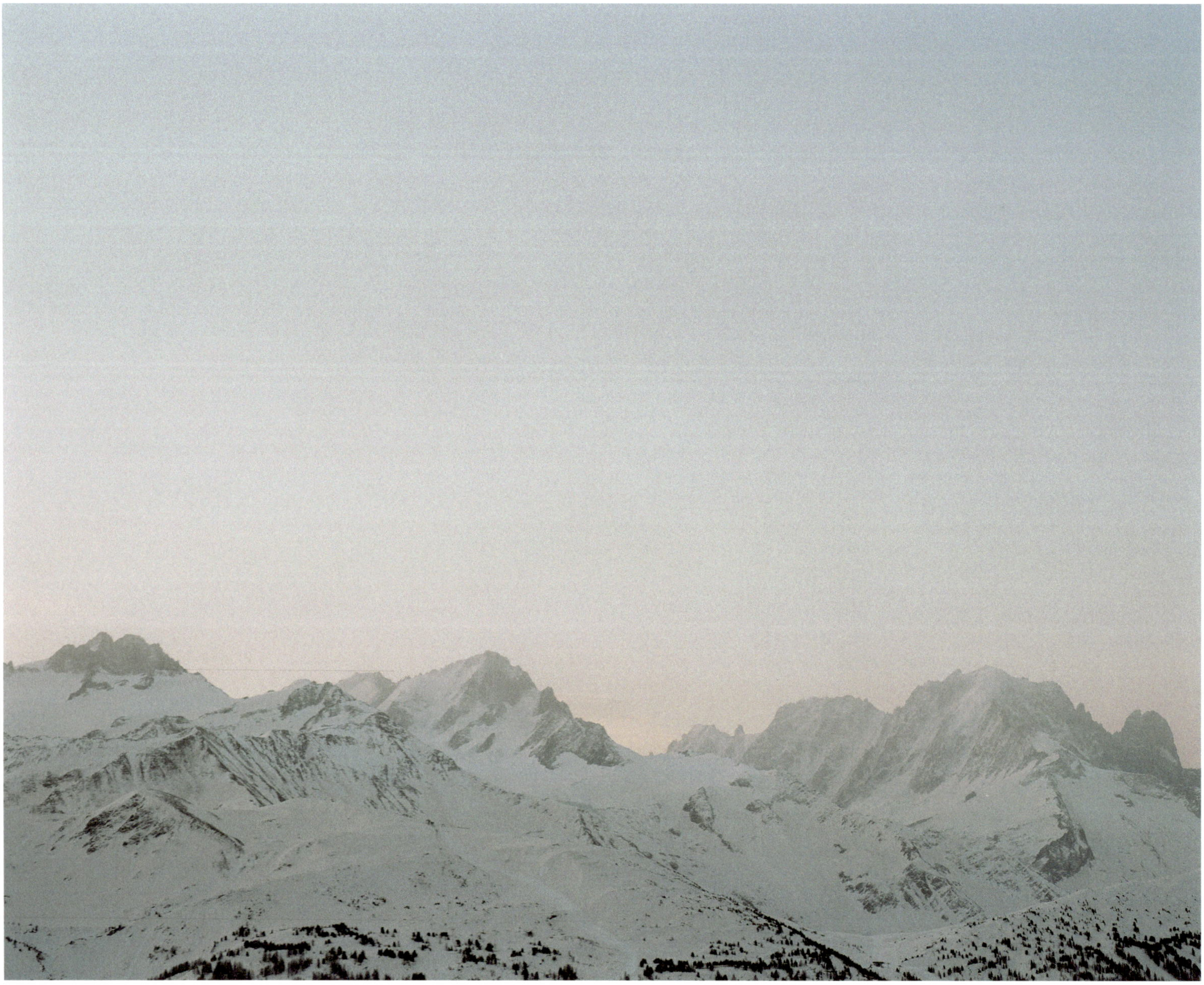

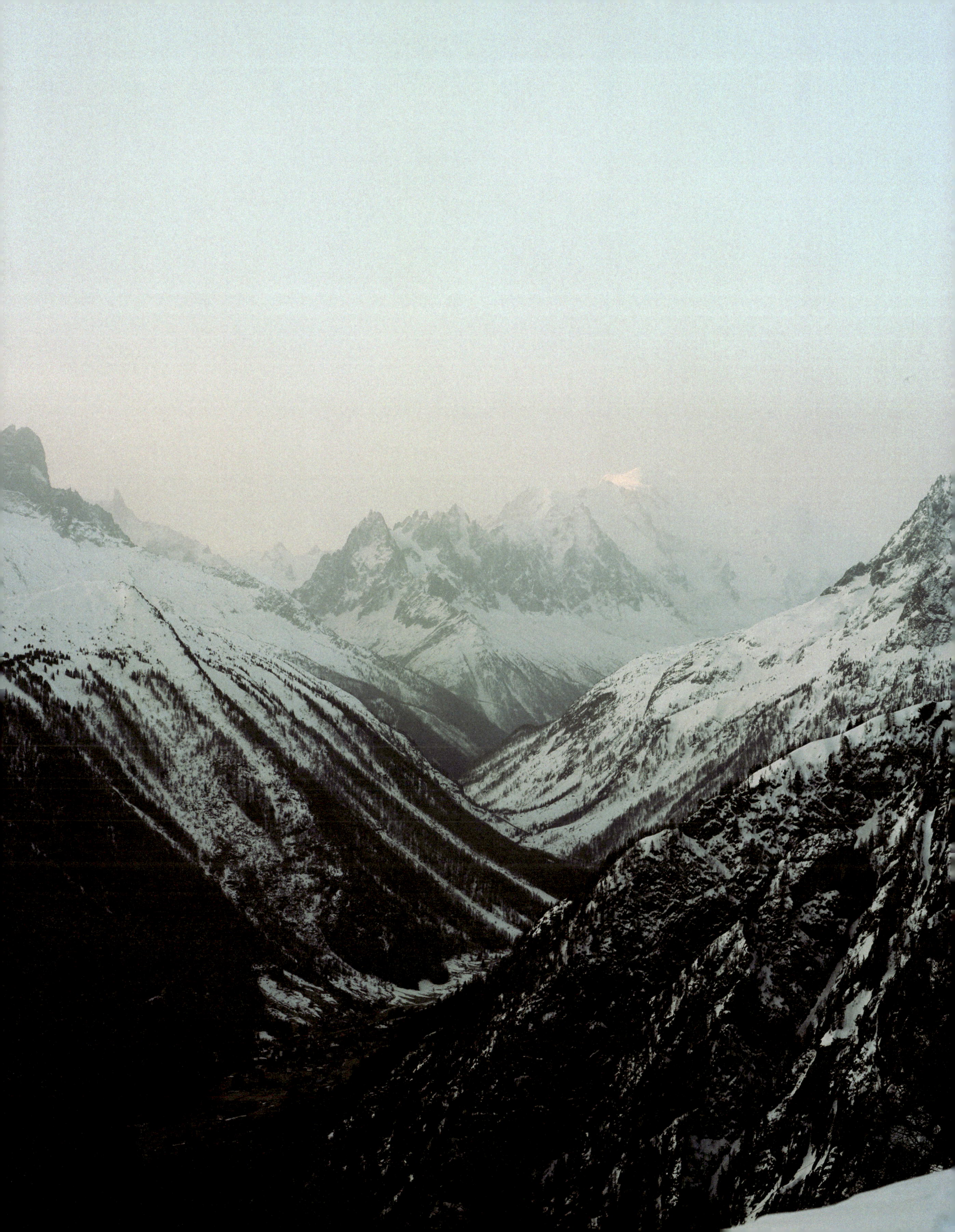

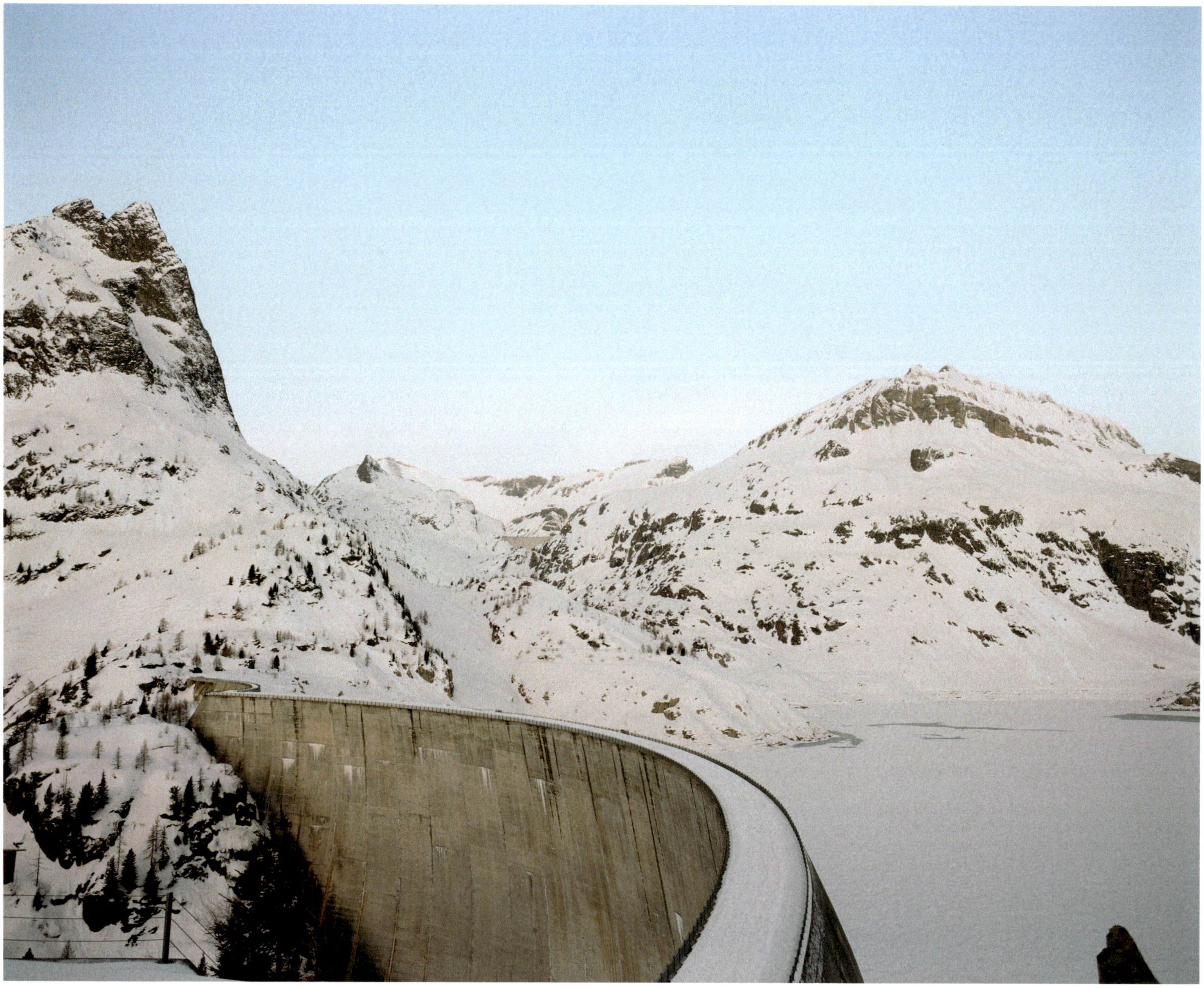

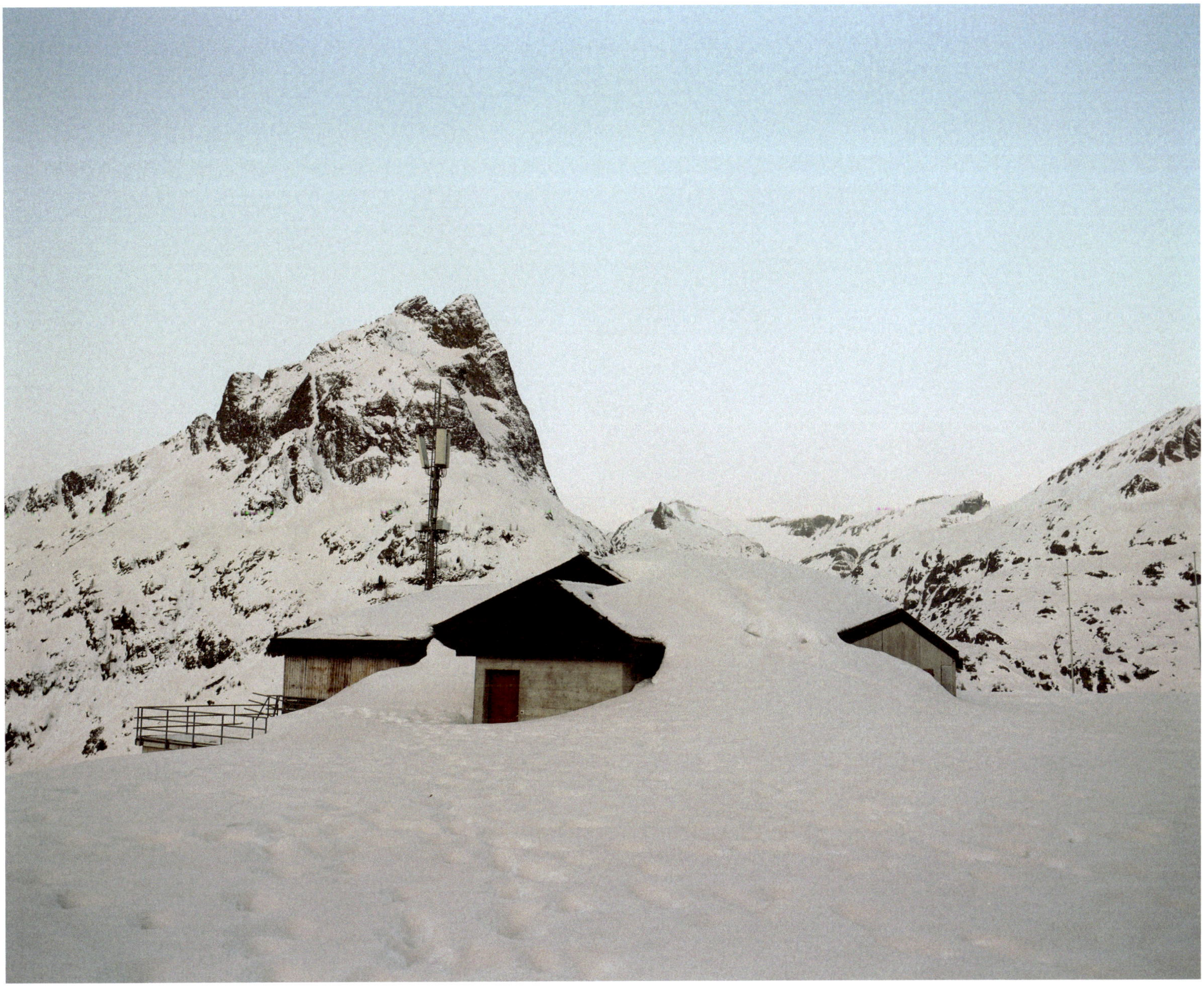

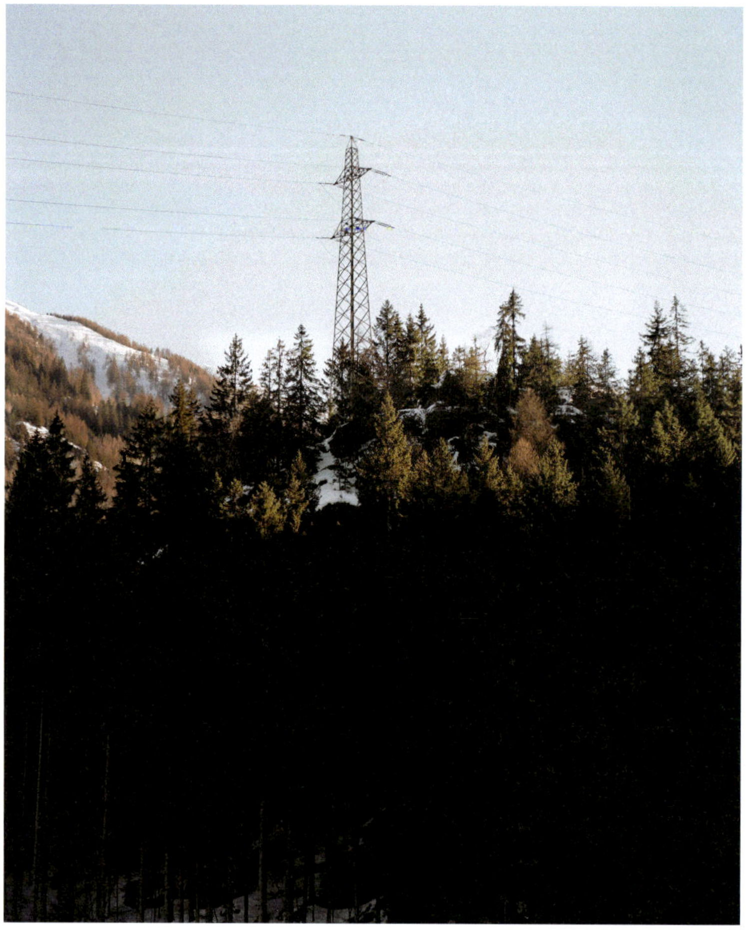

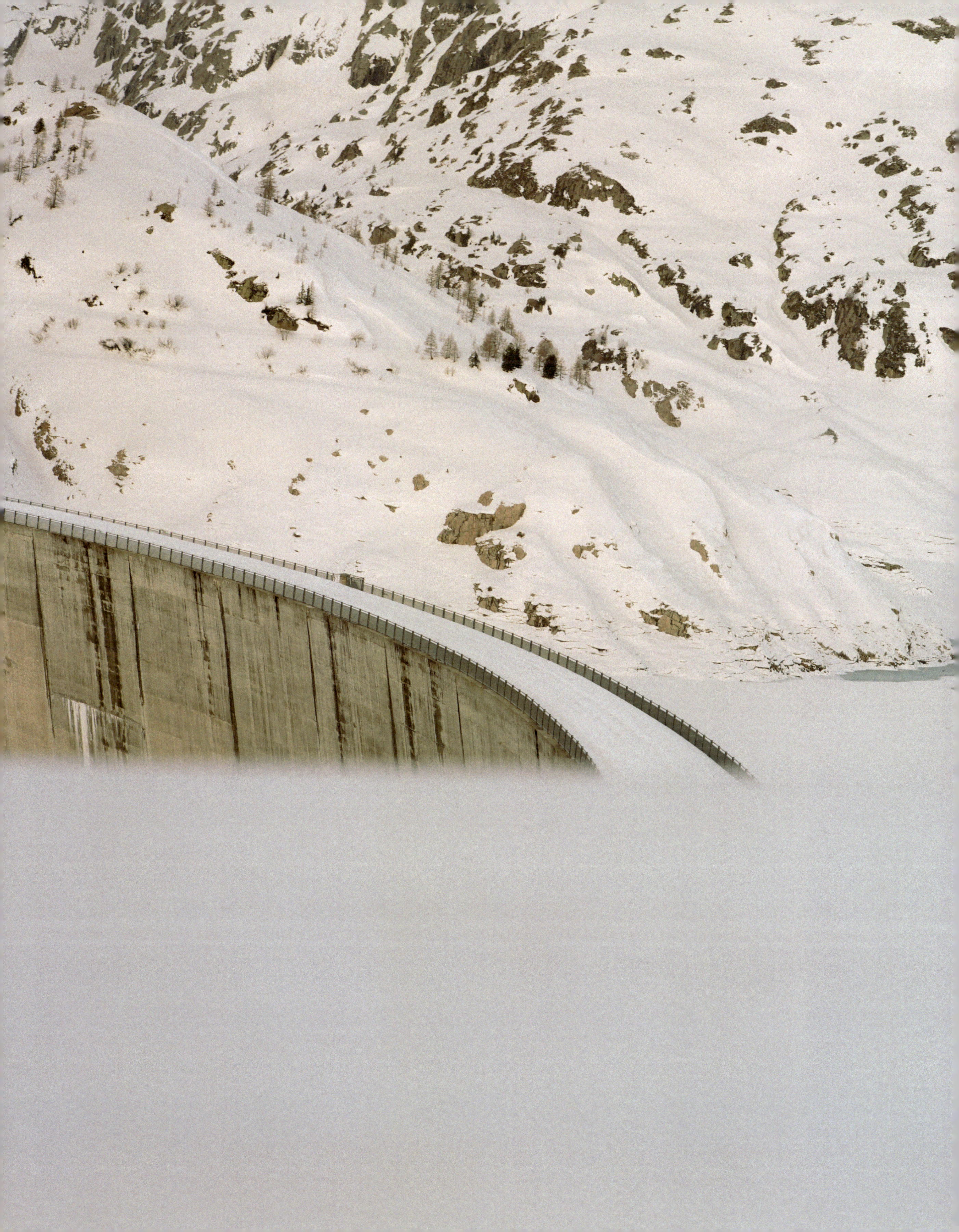

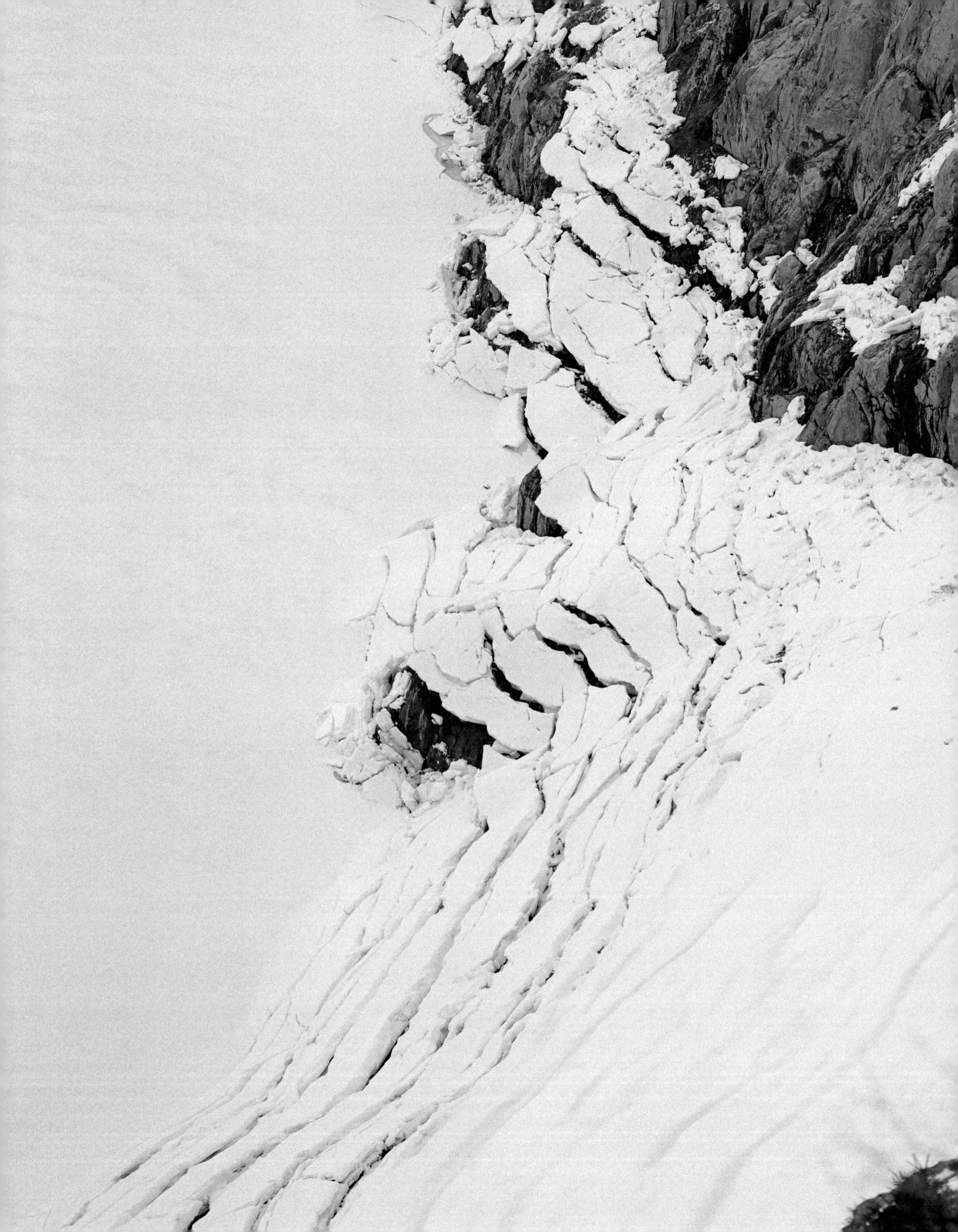

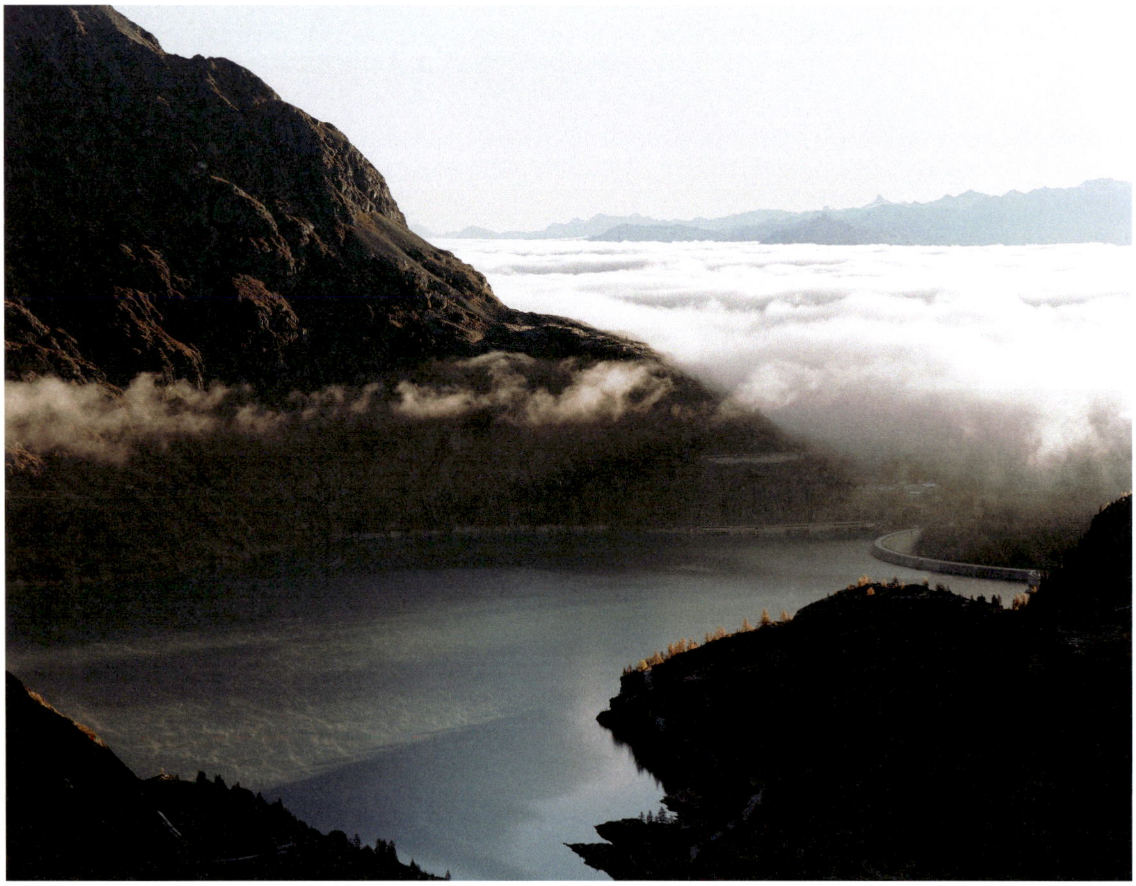

No word is more difficult to define than nature, even though everyone uses it frequently. Since ancient times, philosophers have also struggled with a definition. For the moment, however, it is probably easiest to define nature as something that is not an "artefact," something not made by human beings, something that is an opposing concept.

The Emosson dam is precisely where we encounter a sharp confrontation between our archetypal concept of "nature" (the Alps) and the unmistakable "artefact" (concrete structures). Whereas today we can without hesitation call the towering peaks of the Alps things of natural beauty, before the eighteenth century people did not call the majestic mountains "beautiful," as they were beyond the reach of the intellect's perception and understanding of order and harmony. The English who made the Grand Tour in the eighteenth century gave an appropriate adjective to this majestic nature: "sublime," something quite different from beauty. The sublime is an aesthetic category that describes the feelings one has towards something enormous and terrible, far beyond the human scale. However, the term was initially used for spiritual activities such as heroic endeavors. When we stand in front of this magnificent landscape, we are surrounded by mixed emotions. In front of this sublime landscape of the Alps, we feel awe, and at the same time, we feel the same kind of emotion and admiration for the long, massive concrete walls that man has built there. This is because, while the towering Alps are magnificent beyond the human scale, the dam embankments show that humans have altered and exploited nature.

For some people, the arched concrete structures created in the natural environment may remind them of the earthworks created in the American wilderness in the late 1960s and 1970s. This is not surprising because, although the term “earthwork” was coined to mean works using soil, it was originally used to describe civil engineering works, and this type of environmental art form, created in the landscape using the landscape itself as material, was reminiscent of such works.

The eighteenth-century philosopher Immanuel Kant wrote about the “inaccessibility of nature” as “nature is sublime in those of its appearances whose contemplation carries with it the idea of its infinity,”[1] but clearly since the twentieth century civil engineering technology has accessed, divided, and exploited this Alpine nature by works. Discussions of environmental ethics are often framed in terms of a dichotomy between instrumental anthropocentrism, which sees nature as having value as a tool available to humans, and intrinsic physiocentrism/ ecocentrism, which emphasizes the intrinsic value of nature. More to the point, nature has been considered undesirable from a traditional conservation perspective, which values nature in terms of its instrumental usefulness. However, the contemporary natural environment is a complex intertwining of natural and artificial factors, which makes it extremely difficult to present simple pros and cons and instead requires a pluralistic judgment.

After the accident at the Fukushima nuclear power plant in 2011, a better energy option, one that was safer and more efficient, with the least environmental impact, was required. The construction of the Nant de Drance Hydropower Plant, a new pumped storage power plant to re-use the dammed lakes of two existing dams, the Emosson dam from 1973 and the even older, upstream Vieux Emosson dam from 1955, involved the construction of a vast water pipeline connecting the two dammed lakes; but thanks to twenty-first-century civil technology, the entire power station was built underground in the Alps so as not to spoil the natural landscape, while at the same time preserving the lost natural environment. A pumped storage power plant operating with the benefit of Alpine water appears to be a natural storage battery and a practical and sustainable way of eliminating the waste of electricity generation, which is usually challenging to control. Water is one of the four elements of nature that is indispensable for sustaining life. That is why we may have had the somewhat unwarranted belief that water is a perpetually circulating entity that will continue to provide us with blessings. Ironically, however, the glaciers in the Alps, which are the source for the Emosson dam, have been particularly affected by global warming and, in the worst-case scenario, are even predicted to almost disappear in the next hundred years.

1 Immanuel Kant, “§ 26 Von der Grössenschätzung der Naturdinge, die zur Idee des Erhabenen erforderlich ist,” in *Kritik der Urteilskraft* (Berlin/Liepāja: Lagarde und Friedrich, 1790), 93; (Akademie-Ausgabe) 206. English translation by the author.

Against the backdrop of the Alpine mountains with their white snow and ice, Emosson Lake is filled to the brim with the deep blue water that is typical of dammed lakes, reflecting the peaks on its surface and contrasting vividly with the grayish-white, gently arched concrete walls. The outstanding scenic, sometimes breathtakingly beautiful sight above the clouds is clearly a natural environment containing an artificial dam. Consider that we live in the twenty-first century and are not so concerned about the intervention of artificial objects in nature. Concrete dams have a relatively long history, as they were built as part of modernization. Moreover, the concept of "untouched nature" is already an illusion, mainly since the Alpine landscape is no longer an unexplored one exempt from human activity but one that has been surveyed and often altered. The steep mountains of the Alps, once a terrifying presence, have become a place for people to enjoy aesthetically, and the Emosson dam has also become one of the leading tourist destinations.

However, in the age of climate change, with the loss of so much of the natural environment, will it become impossible for people to enjoy the environment aesthetically? Many tourist attractions and everyday places once considered beautiful may cease to exist, or we may only perceive negative aesthetic values, plagued by moral responsibility and guilt. Indeed, we are increasingly seeing warnings that all the effects of climate change are so intricately intertwined and so devastatingly destructive to the environment on a global scale that, if nothing is done, the planet's life-support systems will collapse. The survival of human civilization will be at stake.

When the water in the dams is depleted, and humans are gone, the concrete structures will remain. Even if they remain as relics that retain traces of the human activity (= art) that once existed there, no one will be present to contemplate or give meaning to the artificial objects. At that time, nature may once again surpass everything with its inaccessibility. The natural landscape of the not-too-distant future that the Emosson dam will depict will soon be tinged with sadness at the loss of water and the world that once existed.

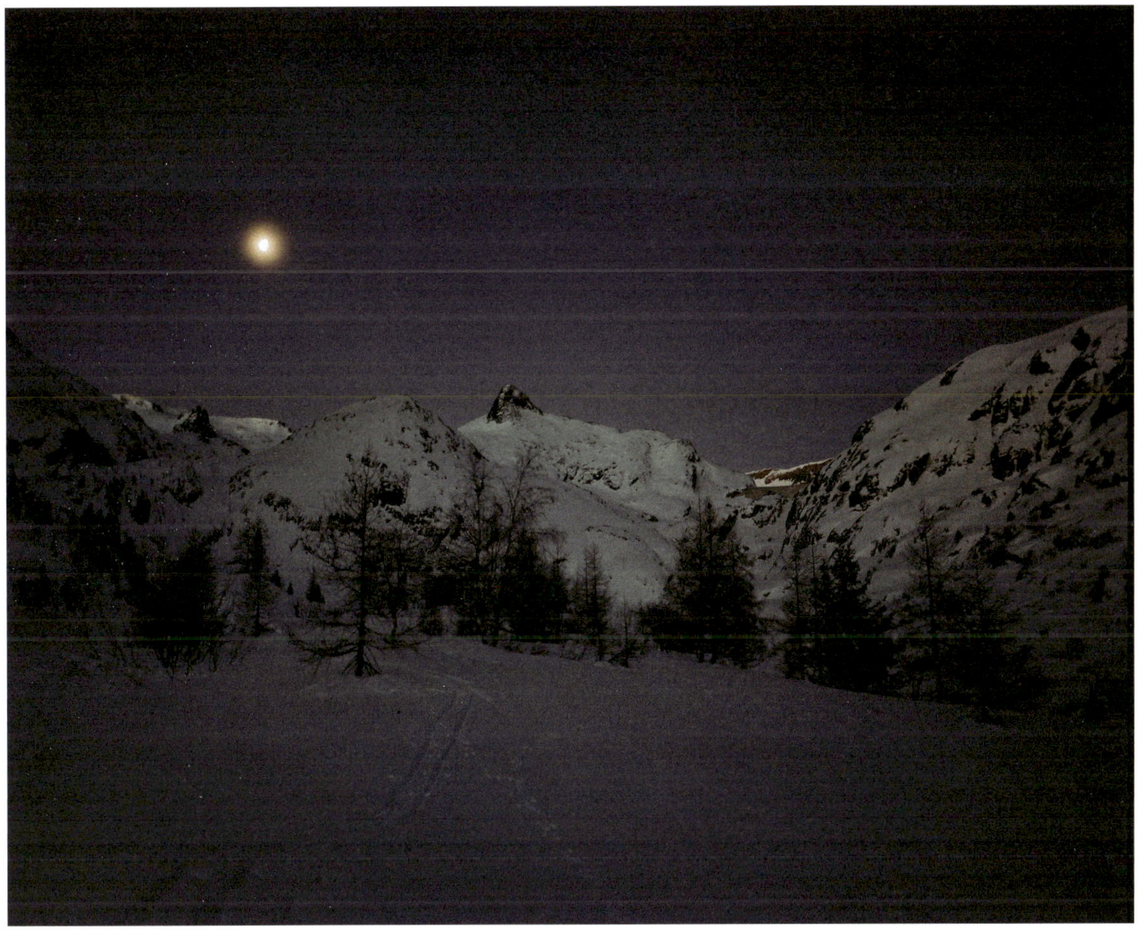

誰もが普通に口にするのにも関わらず自然ほど定義の難しい言葉はない。古代から哲学者も「自然」の定義に苦慮してきた。しかし、差し当たっては対立概念である「人工」ではないもの、という定義が最も誤解がないかもしれない。

エモッソン・ダムは、まさに、私たちが抱く典型概念としての「自然」（＝アルプス）とこれまた間違いようのない「人工」（＝コンクリートの建造物）が尖鋭的に対立している現場に出会える場所である。

現代の私たちは、躊躇なくアルプスの屹立する山々を自然美と呼ぶことができるが、18世紀以前、秩序や調和といった知性で捉えられる範囲を超えた雄大な山々を美と呼ぶことはなかった。18世紀にグランド・ツアーを行ったイギリス人たちは、この雄大な自然に相応しい形容詞を美とはまったく異なる「崇高（sublime）」という概念で表現した。崇高とは人間のスケールを遥かに超えた巨大なものや恐ろしいものに対して抱く感情を表す美的範疇であるが、もともとは英雄的な行為のような精神的な営みに用いられる言葉だった。このアルプスの崇高な風景の前で、わたしたちは畏怖にも似た感情を抱き、同時にそこに人間が築いたコンクリートの長く巨大な壁にも同種の感情を抱き、感嘆の声をあげる。この雄大な風景を前に立ち尽くすとき、複雑な感情に包まれる。というのは、アルプスの屹立する山々は、明らかに人間のスケールを凌駕する壮大さである一方で、ダムの堤体が示すのは、人間の手によって自然を改変し利用しているという事実だからである。ひとによっては、60年代後半から70年代にアメリカの荒野に制作されたアースワークを連想するかもしれない。それが頷けるのは、「アースワーク」という言葉は土を用いた作品という意味での造語ではあるが、もともと土木工事を表す言葉であり、風景の中に風景そのものを素材にしてつくられるこの種の環境芸術の形態が土木工事を想起させるものだったからである。

18世紀の哲学者カントは「自然が、その現象において崇高であるのは、その現象の直観が自然の無限性という理念をともなう場合である」[1]として、「自然の近づき難さ」について語ったが、20世紀以降の土木技術は明らかにこのアルプスの自然に近づき、分け入り、土木工事を施すことで利用している。環境倫理学の議論では、しばしば自然を人間にとって利用可能な道具として価値を持つと考える道具的な人間中心主義と、自然自体が固有に持つ内在的な価値を重視する内在的な自然中心主義の二項対立で捉えられることが多い。さらに言えば、伝統的な自然保護の観点からすると自然を道具のような有用性で価値づけすることは好ましくないとされてきた。しかし、現代の自然環境は自然と人工が複雑に絡み合い、単純な是非を問うことがきわめて難しく、多元的な判断が必要とされる。

気候変動の大きな要因とされるCO2の削減が喫緊の課題であっても、電力の使用を否定することは今日の私たちには不可能であろう。2011年の福島原子力発電所の事故以来、より安全かつ現時点で最も効率が良く、環境負荷を最低限に抑えられる最良の選択が求められている。二つの既存のダム、1973年にできたエモッソン・ダムと、さらに古い、1955年にできた上流のヴュー・エモッソン・ダムのダム湖を再利用する新たな揚水発電所「ナント・ド・ドランス発電所」の建設は、21世紀の土木技術によって、巨大な導水管の建設を伴ってはいたが、自然の風景を損なうことがないようにアルプスの地下に建設し、同時に失われる自然環境を保全するための工事も忘れなかった。アルプスの水の恩恵を受けて稼働する揚水発電所は自然の蓄電池となり、コントロールの難しい発電量の無駄を無くす、現実的で持続可能な方法に見える。水は、自然の四大元素の中でも生命を維持するために欠かすことができない。だからこそ、水が永遠に循環し私たちに恵みを与え続ける存在だと、どこか無根拠な信念を私たちは持っていたのかもしれない。しかし皮肉なことに、エモッソン・ダムの水源であるアルプスの氷河は、地球温暖化の影響がとりわけ顕著であり、最悪の場合、100年後にほぼ消滅するという予測すらある。

1 イマヌエル・カント『判断力批判』熊野純彦訳 作品社 2015 p.196

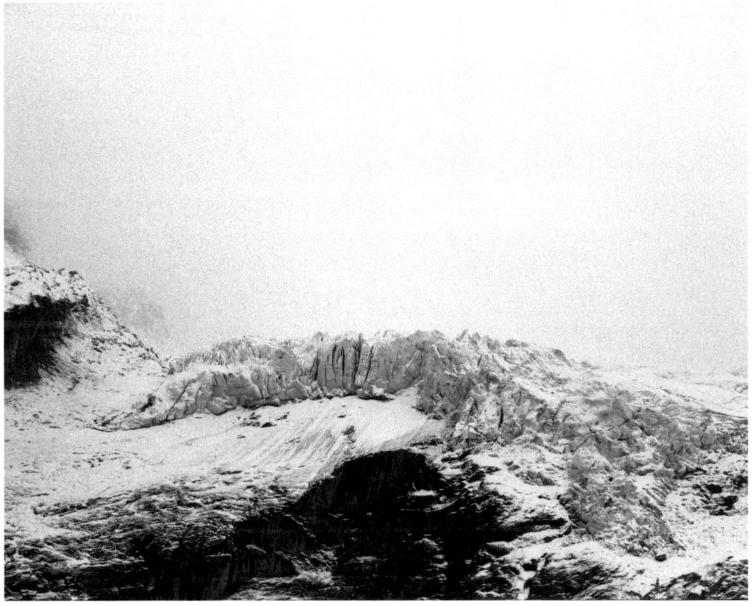

白い雪氷を抱くアルプスの山々を背景に、ダム湖特有の濃い青色の水をなみなみと湛えたエモッソン湖は、湖面に峰々を映し取り、灰白色の緩やかにアーチを描くコンクリートの壁と鮮やかなコントラストをなしている。風光明媚な、雲上に広がるときに息を呑むほど美しいその光景は、明らかに、ダムという人工を含み込んだ自然環境である。思えば、21世紀に生きる私たちは、自然の中にある人工的なものの介入をそれほど気にしない。なかでもダムの歴史は近代化と共にあり、かなり長い。また、「手付かずの自然」という概念がすでに幻想になっている今、とりわけアルプスの風景は、それを眺める私たちの視点を獲得して以降は、人の営みと離れた存在ではなくなっている。かつては恐れられる存在だったアルプスの急峻な山々は、人々に美的に観照され、享受される場所であり、エモッソン・ダムもまた、有数な観光地の一つになっている。しかし、気候変動の時代において、現実に多くの自然環境が失われていく中で、人々は環境を美的に楽しむことが不可能になってしまうのだろうか。かつて美しいとされた多くの観光地や日常的な場所が存在しなくなるか、あるいは私たちは道徳的な責任や罪悪感に苛まれた負の美的価値しか感じられなくなるかもしれない。実際、気候変動による全ての影響は複雑に絡み合い、破滅的な環境破壊を世界規模でもたらすため、このままでは地球の生命維持システムが崩壊し、人類の文明の存続も危ぶまれるという警告を目にすることも多くなっている。

ダムの水が枯渇し、人間がいなくなった時も、コンクリートの建造物は残る。かつてそこにあった人間の営み＝技術（art）の痕跡をとどめる遺跡としてそれが残ったとしても、眺める存在を失った人工物に意味を与える存在はもうそこにはいない。そのときふたたび、自然は近づき難さを持ってすべてを凌駕するのかもしれない。エモッソン・ダムが描いてみせるそう遠くない将来の自然の風景は、やがて失われる水とかつてあった世界への悲しみを帯びることになる。

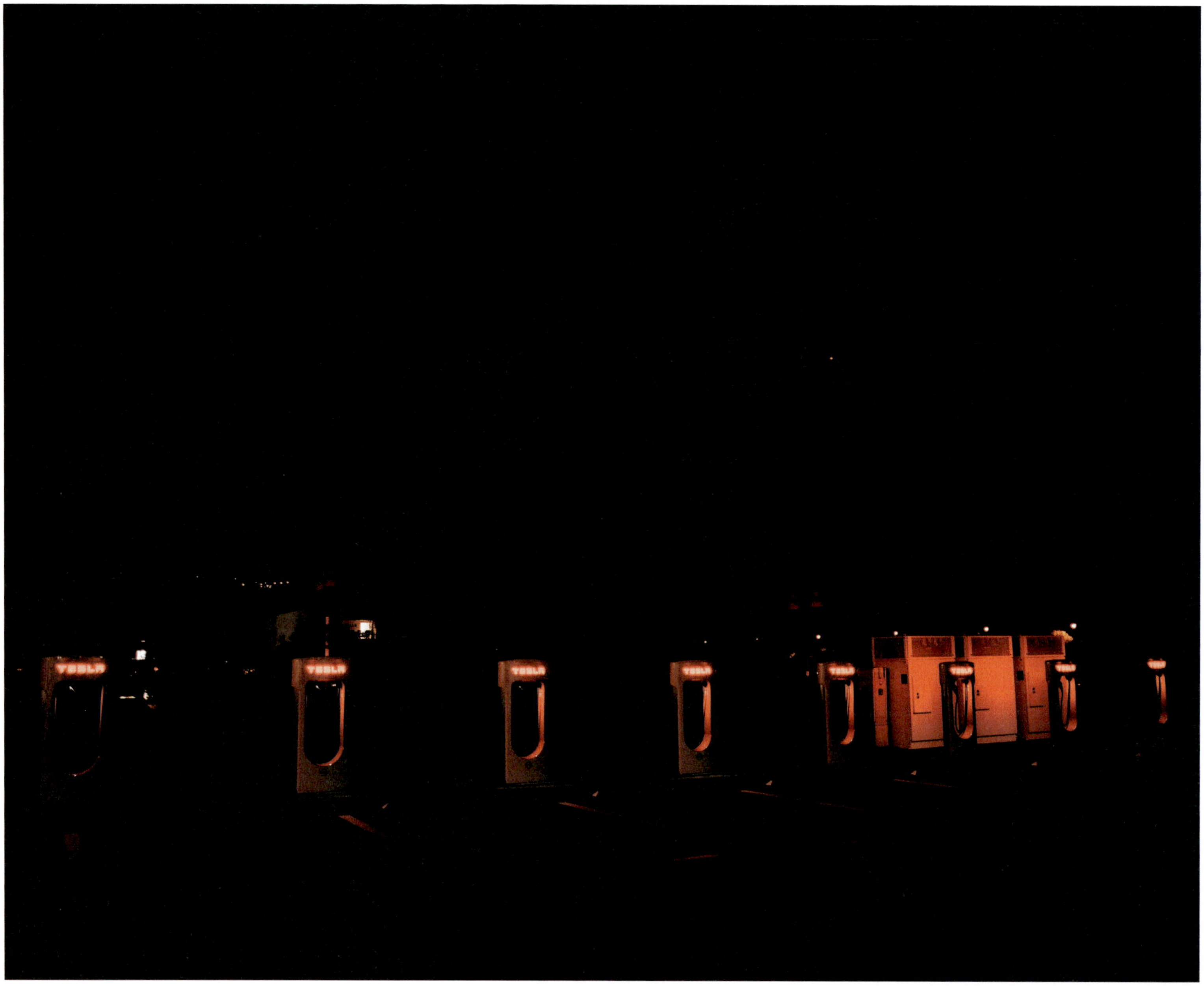

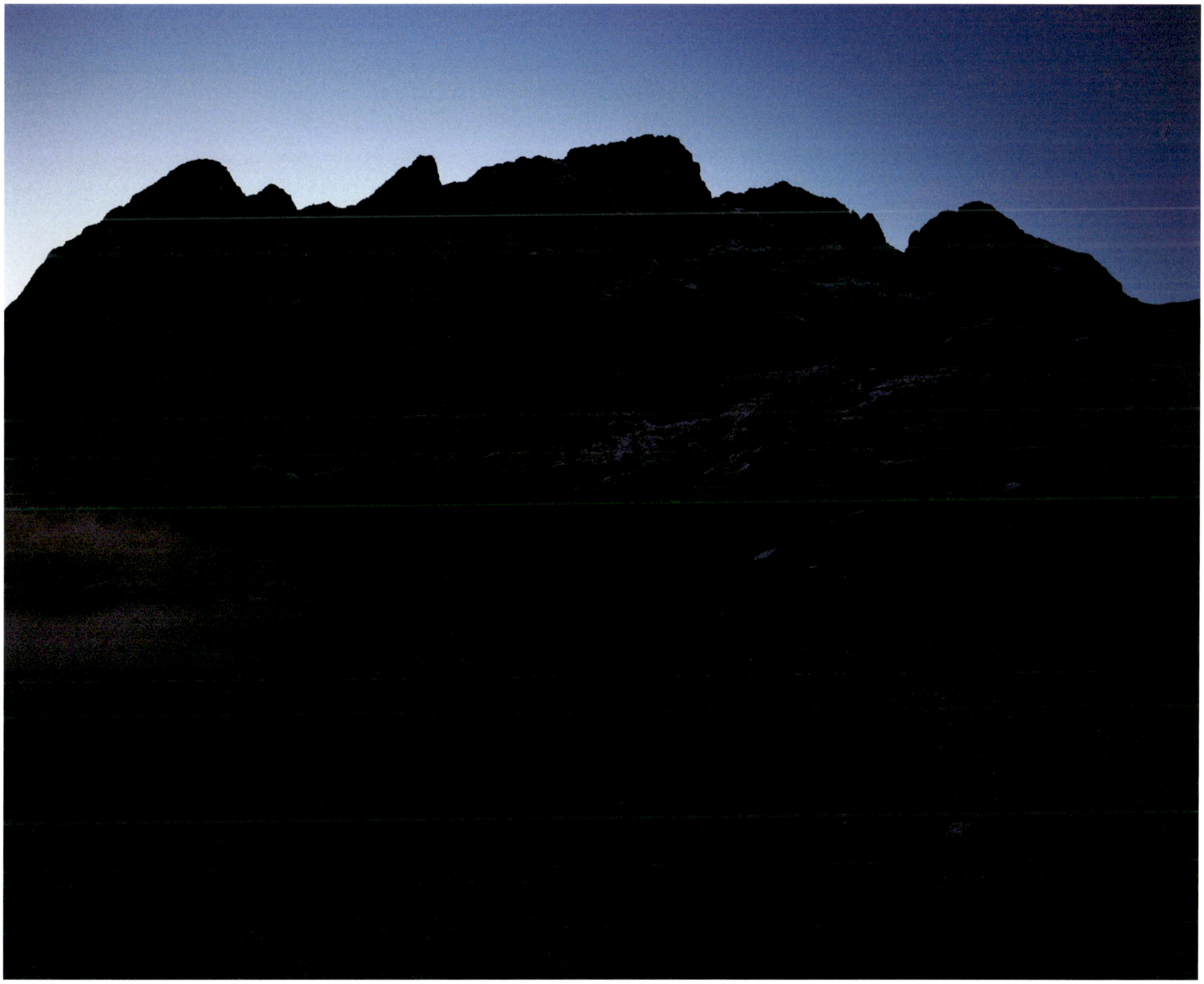

The majestic construction of the dam, along with its eternal embrace with nature, stands as a tangible symbol of human resilience throughout time. It will remain firm and immobile, despite the water pressing against its concrete shoulders, as if foreshadowing the creation of a “pyramid” in the Swiss Alps—an awe-inspiring monument erected by human hands amidst nature. This structure testifies to human intelligence, representing a magnificent engineering and architectural intervention awaiting future advancements for energy transformation and conservation: indispensable batteries for our way of life that, woe betide us, will otherwise descend into blackout.

Film photography, with its meditative rhythm and limited number of shots, has taught us patience and awareness of the present moment. We embarked on our journey with empty rolls, and returned with them filled. Alongside our equipment, we carried the lighter weight of emotions and memories captured on those fragile negatives.

We traversed the seasons, ascending through fields abuzz with bees and inhabited by rational ibex, and descended navigating recent avalanches.

Upon our arrival, we found it challenging to assign a non-engineering meaning to the beginning and end of the dam. When viewed from below, the dam’s imposing presence is revealed, with its concrete walls reaching skyward. It is a tangible starting point from which the water’s energy accumulates and prepares to flow. However, upon reaching the summit, we perceive the water’s force being held back by these walls, and find ourselves at the precipice of the pressure it exerts, as if it were the inception of an unstoppable force. Our perception expands, and we realize that beginnings and endings are relative concepts, dependent on our point of view and perspective.

As frigid nights enveloped us, the mountain’s breath seemed to synchronize with our own. In that place, far removed from the frenzy of daily life, we learned the value of patience and contemplation. Sleeping in a cold and dark environment, near the basin, we could hear occasional sounds of ice breaking on the dam lake, accompanied by extraterrestrial echoes from the depths of the dam.

The darkness of the sky and the silence were absorbed by the blanket of snow that enveloped everything.

果てしなく広がる自然の中に造られるダムという壮大な建築物は、いつの時代も人類の底力を表す有形のシンボルとして存在している。コンクリートの堤体にかかる水圧にもかかわらず、頑丈かつ不動なものにしなければならない点では、スイスのアルプスという大自然の中に人間の手で建造する荘厳なモニュメント、まるで「ピラミッド」の建築を予感させるものだった。この建造物は、将来的にエネルギーの転換と保存の実現を約束する偉大なる技術と建築の介入の表れであり、人間の英知を証明するものだ。さもなくば、災いなるかな、停電の憂き目に遭いかねない私たちの生活にとって、必要不可欠な電力源なのだ。

フィルム写真は、その瞑想的なリズムと限られた撮影枚数で、私たちに忍耐と今この瞬間を認識することを教えてくれた。空っぽのフィルムと共に出発して、写真でいっぱいのフィルムと共に帰る旅だった。撮影機材と一緒に運んだのは、繊細なネガに収められた、物理的な重さのない感情や記憶だ。

私たちはいくつかの季節をこの山で過ごした。ミツバチが飛び交い、穏やかで賢いアイベックス（ヤギの一種）の棲む野を抜けて山を登り、雪崩の跡を避けながら山を下ったのだ。

到着してみると、ダムの始まりと終わりに工学以外の意味を持たせることに困難を感じた。下からダムを見上げると、空へと伸びるコンクリート壁の威風堂々たる存在感は圧巻で、水のエネルギーを蓄積して放出する原点を目の当たりにした。ところが、ダムの上に着くと、ダム壁で水の力が堰き止められているのが分かる。私たちは、水の圧力を受け止めるダムの絶壁に立っており、そこはまるで止めようのない力がみなぎる原点のようだった。私たちの知覚は拡大し、始まりや終わりとは、自分たちの見方や視点によって決まる相対的概念なのだと悟った。

極寒の夜に包まれると、山の呼吸と自分たちの呼吸がシンクロしているように感じた。狂乱した日常生活からかけ離れたこの場所で、私たちは忍耐と思索を学んだ。寒くて暗いなか、ダム湖の近くで眠ると、時折、氷が砕ける音と共にダムの底から異星人のこだまが聞こえる。

空の闇と静寂が、すべてを包み込む一面の雪に吸い込まれていく。

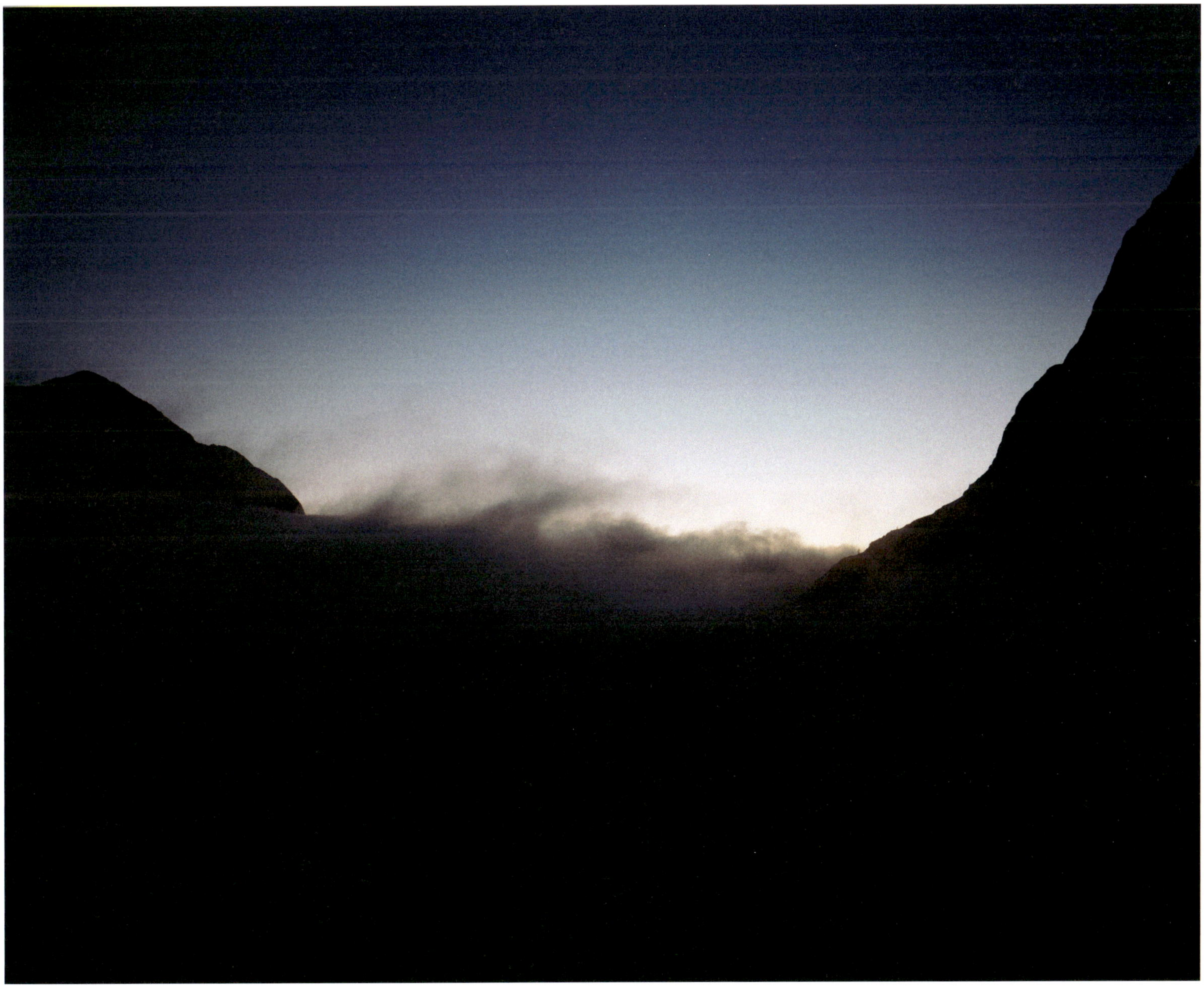

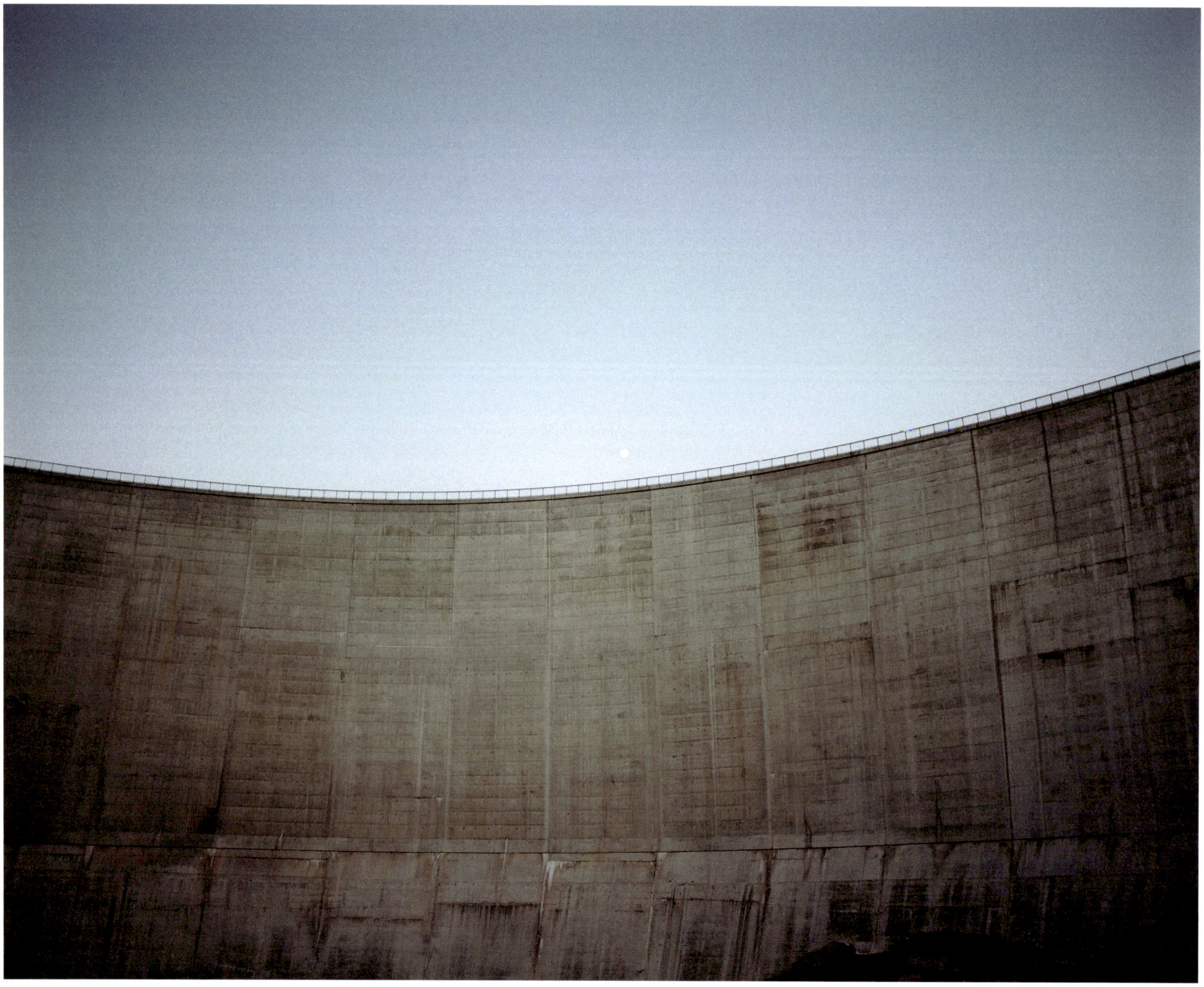